An aethereal theory of gravity

Peter Mason Atkinson

Summary – The theory in a nutshell

Let me try to really simplify the theory so that anybody can understand it. Point the nozzle of a vacuum cleaner at a tiny piece of paper on the floor. As the air accelerates into the vacuum cleaner the air sucks the paper along with it into the machine. Substitute the earth for the electric motor, substitute the aether for the air, and substitute yourself for the piece of paper. The earth sucks in and destroys aether, and as the aether rushes in you are caught up in the accelerating air flow, and that is why you stick to the earth. That's it. That's gravity. It really is very simple.

Abstract

A new theory of gravity is presented which is based on the existence of the aether. The history of the aether is discussed and the erroneous reasons why it was abandoned are identified. The wave particle duality of light is discussed and a solution for this problem is offered. An explanation for straight edge diffraction is given. The observation of Galileo that all objects fall at the same speed is explained in detail. This theory of gravity is that aether is destroyed in all matter, and as it is destroyed, aether moves into the object to keep the density of the aether constant. As the aether accelerates into the object other particles are dragged along with the aether and this is the cause of gravity. The reason why the speed of light in a vacuum is a constant is explained. The speed of light is limited by the speed of the aether compression waves that surround moving photons and the speed of these in turn dependent of the density of the aether. The density of the aether is a constant therefore the speed of light is a constant. The fundamental laws of aether dynamics are given and the principles of aether flow are discussed. The concept of drag forces and impact collisions are clarified. Gravity is not an attraction, it is a drag force. Using the principles of aether dynamics the concepts of mass, energy and momentum are explained. Using aether dynamics Newton's law of gravity is explained, as are his three laws of motion. A new universal constant is identified. The mass of aether in Kg destroyed by one Kg of matter in one second is a universal constant. The reason for the failure of Michelson's and Morley's experiment is discussed. Using this theory a relationship between the orbital velocities of the planets and the distances to the sun is derived. The theory is used to demonstrate that the center of the galaxy is a black hole. The accelerating expansion of the universe is explained without the need for dark energy and Olbers's paradox is solved. It is demonstrated that the recession of the moon is due to decreasing aether density and the recession of the moon obeys Hubble's law. It is proposed that the visible universe is only one part of a vastly larger universe that can never be seen. A relationship between gravity and the strong nuclear force is proposed. An aether compression wave is proposed as a possible mechanism of transmitting supernatural forces. A possible way of proving this gravity theory using heavy water is given.

Preface

Who is this book written for? It is written for the amateur astronomer, and the intelligent man in the street. Seeing everyone on the street is rather more intelligent than the average, it is written for everyone. I have tried to remove nearly all mathematics, and what is left is very simple, and easy to understand. If this theory cannot be understood by a school leaver then I have failed in my task. I have also tried to make it interesting and read easily. This subject could easily become very difficult and extremely boring, hence my rather unorthodox approach.

Introduction

For thousands of years people have been asking the question. Why do things fall to earth? Many of the world's greatest scientists including Galileo Galilei (1564-1642), Isaac Newton (1642-1727) and Albert Einstein (1879-1955) have addressed this question. Galileo and Newton were shrewd enough not to offer an explanation of what gravity actually is. Einstein bravely offered a theory of what is actually causing gravity. Newton's concept of gravity lasted for about 250 years, and Einstein's has been in fashion for about 90 years. Thousand of scientists have written papers and spend lifetimes working with these theories and equations. Both Einstein and Newton came up with equations, which accurately predicted the behavior of planets and other heavenly bodies. Einstein's equations are somewhat more accurate than Newton's. But as far as what is the true cause of gravity we actually have only one theory that has been accepted by the scientific community and that is Einstein's theory that gravity is due the warping of space time by objects. While everyone is happy to accept that his equations work for planets, not everyone is happy with his explanation of the cause of gravity, especially as it seems to breaks down at the subatomic particle level. Many other theories have been put forward by many people, but none have gained any recognition by the scientific community. I postulate that both Newton and Einstein were fundamentally wrong about the true cause of gravity and will present an entirely different theory, based on a set of entirely different concepts about matter and the universe.

I believe that over the last hundred years a few serious mistakes have occurred in scientific understanding that have lead the scientific community astray, not only regarding gravity, but all the other fields of physics as well. These errors were due to the incorrect interpretation of experimental results. The misinterpretations were made because of incomplete understanding of the nature of the universe. At the time the current theories of gravity were made very little was known about the universe. Indeed when Einstein was proposing his special and general theories of relativity, galaxies had not even been discovered. Our view of the universe today is vastly more detailed than any previous generation could possibly have imagined. The mistakes that were made were in large part due to a lack of information. Nevertheless they have lead to a fundamental misunderstanding of the nature of gravity and all the other natural forces. This paper expounds a theory to explain gravity, and alludes to other natural forces and cheekily proposes a possible rational explanation for a mechanism underlying supernatural phenomena.

It is understandable that most physicists working in the gravity field are reluctant to criticize current theories. Research grants can disappear in a heartbeat, especially if there is any suspicion that the beneficiary is a little odd, or heaven forbid 'anti establishment'. I am in the fortunate position where I don't have to worry about what other people say about my theory. The only risk I take is ego deflation. Another reason why these errors have been perpetuated besides fear of rocking the boat, is that the equations for gravity derived from the theories work very well. If you can use a theory to arrive at accurate figures, can the theory really be wrong? Well in most cases the answer would be no, but in the case of gravity I think the answer is yes. The theory can be wrong but the equations correct. The reason for this is that the equations were derived empirically from observations, and the theory plugged in afterwards.

I think that we all suffer from certain weaknesses of the human brain that lead to systematic errors. In particular there are two flaws. The first I call cerebral filtering and the other humanization. These two flaws prevent us from seeing things as they really are.

Cerebral filtering

This is where the brain shuts out information that cannot be reconciled with other existing knowledge. Let's look at a very famous musical verse.

> Nothing comes from nothing,
> Nothing ever could,
> Somewhere in my youth, or childhood
> I must have done something good.
>
> Sound of music Rodgers and Hammerstein.

Well this is certainly good music but is it true? What about the first line; that nothing comes from nothing. Now I want you to keep an open mind about this. This is important because it is one of the fundamental tenants of my theory. The tenant is that creation is a continuous process and this process causes gravity. If you really think about it you will realize that the statement, nothing comes from nothing simply cannot be true. The universe must have come from nothing. Whether the universe began 15 billion years ago, with a big bang as is the current popular theory, or whether it has existed much longer than that, or whether it came from a parallel universe, or some other exotic origin, it must have come from nothing. Matter just popped out of thin air, well not exactly thin air. But it just popped up out of nothing. Now I know most of will want to close this book just about now, but please don't – there are many more irreverent statements to come.

I believe that the human brain has a mechanism, which prevents us from accepting things that seem to be impossible. Just as the brain can suppress very painful memories, it seems to be able to suppress the 'impossible'. There are many examples of the human brain filtering mechanism at work. If you had stood on a soapbox 100 years ago in Hyde Park and proclaimed with absolute certainty that it was possible to completely destroy the entire city of London, with just a few pounds of metal, nobody would have believed you, and you would have been labeled a crackpot. However after Hiroshima people's responses to the same individual would be: - well everybody knows that, tell us something new. Similarly if you had said that all the matter of the earth could be compressed into the size of a bowling ball you would have been ridiculed. In the Middle Ages if you had said that there were at least 100 billion galaxies each with a 100 billion stars, and that the earth was so insignificant that it does not really matter at all, you would have been burnt at the stake. Indeed Galileo was forced to recant his great works when he was charged with heresy.

Today we know that many seemingly impossible things are true, yet we still have great difficulty coming to terms with these facts. Einstein's predictions that time slows down, things get shorter, and you get heavier as you go faster, especially as you approach the speed of light, are very difficult to comprehend, because it just does not make any sense from what we experience in our daily lives. Yet most of these phenomena have already been demonstrated, exactly as predicted.

Let's return to matter popping out of nothing. Although you might reject it out of hand, you have to face the fact that it must be true, no matter how unpalatable it might be. Whether matter was created 15 billion years ago, or thousands of billions of years ago, or billions of billions of years ago, or whether it is being continuously created, at some point, somewhere, it must have come from nothing. It is no good using a religious excuse, saying that God created the universe, because the obvious retort to this is that God must also have also come from nothing. We cannot comprehend how this could be possible. Just as we cannot really comprehend the power of a hydrogen bomb, or that matter is mostly empty space, or the vast size of the universe, or how utterly insignificant we are - we cannot comprehend that matter can come from nothing. Our brains step in and tell us this cannot possibly be true. After you have really thought about this, you will come to believe like me that the universe must have come from nothing. Thinking the opposite does not make sense. Changing the musical line to; - everything comes from nothing, everything always did – might be more accurate but I guess we will just have to leave it as it is. If we corrected it, it would not rhyme and this would upset our brains even more. Lines in music must rhyme – don't they?

Others have also espoused this idea that matter comes from nothing, most notably Fred Hoyle (1915-2001) the famous British astronomer. He proposed that matter was created from nothing in space. He also showed that the creation of only a few atoms per cubic mm would be enough to account for the, then new observation that the universe was expanding. This idea was met by a lot of criticism. Hoyle expanded on the previous ideas of James Jean (1877-1946) that the universe was static. Both these scientists were of the opinion that the universe had no beginning and will have no end. Fred Hoyle was the person who coined the term Big Bang during a radio broadcast in 1949. He was at the time criticizing the new theory of George Lemaitre (1894-1966) that the universe began at one instant in time and has been expanding ever since. Fred Hoyle later went on to demonstrate how heavy elements could be synthesized within stars. At the time of his death his steady state theory of the universe was just about dead in the water. New evidence supporting the big bang theory, such as the cosmic microwave background, had been discovered. This had been predicated before it was discovered. Because of this it has been taken as strong evidence that the big bang theory is correct and Hoyle was wrong. Nevertheless Hoyle was no fool and I believe that his contention that matter comes from nothing is correct.

If matter comes from nothing then the converse is probably also true, that matter can disappear completely from the universe leaving absolutely nothing behind. If I suggest to people that matter just pops out of thin air they generally think I am a little strange. But if I say that matter can disappear completely, without leaving any trace, they are far more accepting of this idea. My theory of natural and supernatural forces is based on the proposition that matter is being continuously created and continuously destroyed, and that all the forces of nature, not just gravity, are fundamentally due to this phenomenon.

Now where does matter come from and where does it go? The short answer is that nobody knows. Let's consider what is known. There are two schools of thought on the origin of the universe. The first is that the universe is in a steady state and has existed for a very long time, much as it is now. This is the steady state theory of Hoyle, and the other is that it started at one point in time at one place. The latter theory, known as the big bang theory, is the one in vogue at the moment. The big bang theory states that all the matter in the universe came from a single point source about 14.5 billion years ago, and the universe has been expanding ever since. There are several lines of evidence suggesting that the big bang theory is correct. There are others however, a small minority today, who are of the belief that the steady state theory is more likely to be correct. Neither theory makes any attempt to say how matter was actually created. If the theories on where matter came from are to put it mildly, rather vague, there is however some experimental evidence that it can disappear. In laboratory-manufactured collisions between matter and antimatter, matter is annihilated, with the simultaneous release of huge amounts of energy. However, energy is released, and matter and energy are interchangeable, as most of the readers reading this theory already know, so matter is not entirely destroyed in this interaction.

For those interested in supernatural phenomena cerebral filtering is a big problem. They cannot reliably demonstrate the existence of these phenomena by conventional means such as photography, sound recordings, magnetic detectors or anything else. The majority of people therefore dismiss them as crackpots. I believe that almost all the cases where psychic phenomena have been 'demonstrated' by photographs and the like are fraudulent, and produced for money. This is tremendously frustrating for the practitioners of the spiritual arts, and up to now there has been no rational explanation for how these phenomena could possibly be mediated. I will attempt to show how phenomena such as mental telepathy and psycho kinesis could actually work.

Humanization - Giving inanimate objects human properties.

I think that, like cerebral filtering, this very human trait is the other thing that lies at the heart of the significant scientific errors in logic made for the last hundred years. This is a very natural tendency that we all have. In some languages inanimate objects are even given their own sex. Even in English we have this. A ship is a she, and a tank would be a him. We even form attachments to inanimate objects. Teenagers fall in love with their cell phones, and even mature adults give inanimate objects magical powers. Just watch the behavior of gamblers. They think that their behavior can alter the number that will result from the throw of a dice. Well maybe it can be influenced, but they are still giving the dice human properties. How many gambling techniques at the roulette wheel have come to grief because people give the wheel the power of memory?

Now what has this got to do with gravity and other forces? My brother pointed me in the correct direction many years ago when he suggested that the current concept of gravity as an attracting force must be incorrect. He pointed out that it is impossible for an inanimate object to attract another object. After this brief Sunday afternoon chat which only lasted a few minutes, like most people would, I initially dismissed the idea that Newton was wrong. Later I realized that my brother must be correct, and have spent the last 40 years on and off trying to solve the problem of gravity. I would love to believe that the initial spark of genius, realizing why Newton must be wrong, came from me, but sadly this is not so. An inanimate object has got no eyes to see where another object is, and no arms with hands to grab it and pull it, and no brain to do the enormously complex calculations required to computer the future position of a moving object. Consider the moon orbiting the earth. The earth cannot 'know' that there is a moon to attract. It would take at the very least a two-way communication to achieve this. The message would have to go out to look for an object. If one were located the message would have to go back to the earth from the moon. This message would have to move much faster than the speed of light or else the earth would lose its "knowledge" of where the moon was. Then when the earth gets the message that there is a moon out there to attract, by the time it arrives the moon has already moved to another place. If you sit down and really think about this you will also realize that the earth cannot possibly attract the moon, or any other object for that matter. Gravity simply cannot be an attractive force. If you generalize this logic to all other forces such as magnetism you will see the same faulty logic in all of these forces. You can rationally work it out that a magnet attracting another magnet is impossible. It certainly looks and feels to us, holding a magnet, like an attraction but this is an illusion. An inanimate object like a magnet cannot possibly know of the existence on another magnet. The magnetic force has to be something else.

Just because something looks quite obvious does not mean that it is true. It looks like the sun goes round the earth, but it does not. The world looks flat, but it is round. Gravity feels like an attraction but it is not. It appears that the north pole of a magnet attracts the south pole, but it does not. The logic that attractive forces cannot exist means that all our understanding of the fundamental forces of nature are wrong, not just gravity. If you are now thinking that I really have lost the plot, consider the possibility that you are just a victim of cerebral filtering, and you are the crackpot not me. After reading this theory I am sure that you will be far more skeptical about the current theory that gravity is an attraction, and you will see the universe through different eyes.

Only two kinds of forces can exist.

I propose that there are only two possible kinds of forces, drag forces and repulsive forces. When I go further into gravity you will see the relevance of this statement more clearly.

A drag force is where an object accelerates in a given direction because of the presence of an accelerating medium. A person being swept away while trying to cross a flooded river might be considered as an example of a drag force, but this would not be entirely correct, as the water is not accelerating. I propose that gravity, magnetism and the nuclear forces holding atoms together are all examples of drag forces. If a particle is accelerating then a drag force is acting it on. Since the medium responsible for a drag force is always accelerating, a particle in the accelerating medium will also be accelerating.

A repulsive force is where an object is in motion because of an impact collision with another object. There are many examples of repulsive forces. The long tail of a comet is caused by high speed particles emitted by the sun, the solar wind, hitting the comet. You might say that this is not correct. What is happening is that the comet is heating up and gas is being released from the interior of the comet that the gas is forming the tail. Well this is true, but why does the gas form a tail, and not just hang around the comet in a cloud? I remember observing Haley's comet with a telescope when it was very far from the sun, and that is exactly what it looked like, just a fussy ball. As it got closer to the sun and the solar wind got much stronger the particles pushed the gas away from the comet and formed the tail. The solar wind of the sun was repelling the comet. The high-speed solar particles were hitting the gas atoms and setting them into motion. As soon as the two colliding particles separate from each other the speed of the impacted particles remains constant. The speed of a photon is constant. The motion of a photon is therefore probably the result of an impact collision between the photon and some particle within the atom, from which the photon has come. If the particle is accelerating we are looking at the effects of a drag force. If the speed of a particle is constant we are looking at the effects of an impact collision.

These two types of forces are the only ones that allow the possibility of one object affecting another object without the first objects having to have prior 'knowledge' about the existence of the other object. All the known forces will eventually be fitted into one of these two kind of forces. This includes gravity, the electrostatic force, the strong nuclear force, the weak nuclear force and magnetism.

I have chosen to consider gravity first because this is the easiest of the natural forces to explain and gives a clearer picture of the concepts and principles involved in all forces. In order to explain gravity I have to discuss in the properties and the scientific history of light. I shall also discuss the old discredited concept of the aether, and like Lazarus I will have to raise the dead, and bring the aether back to life.

The aether

This is spelled in two ways. The older form was spelled with an a – aether, and the later form was ether with no a. I will use the older form as I think it is less confusing as the anesthetic gas, ether, as well as the solvent, petroleum ether, are very well established names in widespread use. Virtually nobody refers to the aether anymore except astronomy historians. This has also been referred to as luminiferous aether as it was first proposed to explain the propagation of light. This has a long history and the term was used in science to denote a medium. From before Isaac Newton's time, until the arrival of Einstein, physicists had speculated that there was an unseen medium through which light, and indeed all matter, moved. This was called the aether. It was supposed to be very hard, and dense, and invisible. It was compared to a very hard crystal. The aether was considered to be very small particles. Light was regarded as a wave traveling through the aether. All the experiments of Newton and others at the time, on light and colour suggested that light was a wave. It was necessary to postulate the existence of the aether because of the wave like properties of light. The rationale was that if sound travels through water and air as waves, which on a superficial examination, appeared very similar to light waves, then light waves must also be traveling though a water like medium.

However there was a big problem when extrapolating this behavior to light, and that was the speed of light. The speed of a propagating wave, like sound, depends on the density and the compressibility of the medium. The higher the density, the faster the wave propagates. The more rigid the medium is, the faster the propagation. The speed of light was first measured by Ole Romer in 1676 by measuring the motion of Io, one of Jupiter's moons. It never ceases to amaze me how much information these early pioneers were able to derive from seemingly insignificant observations using primitive tools. Johannes Kepler (1571-1630) was able to accurately work out the orbits of the planets using the data of Tycho Brahe (1541-1601) who did not even have a telescope. By 1976 the speed of light was known to be 299,792.458 Km/sec. Once it was realized that light travels exceedingly fast, it was also realized that the aether had to be very rigid and incompressible, and basically like a solid. That we lived is an invisible solid was hard for people to swallow, and scientists began looking for an alternative explanation. Now there is an important point to note here. The aether was considered to be a solid because the speed of light was so great, and that light was a wave. For a wave to propagate at that great speed the aether had to be a dense solid. But if light is not a wave, and does not propagate as a wave then the solid aether problem falls away. This has never really been discussed, but the implication is that, if light is not propagating as a wave, then the aether does not have to be that dense.

Diffraction of light

Thomas Young (1773-1829) did numerous detailed experiments with light and in 1803 he demonstrated that when he directed a light beam at two parallel slits an interference pattern was produced. Producing an interference pattern was basically taken as proof that light is a wave. The theories of interference patterns were expanded massively over the next 150 years and the study of interference patterns produces by X-rays passing through DNA crystals was the key that lead to the discovery of the structure of DNA, and how our genetic material actually works. There are many other properties of light that can be explained by assuming that light moves as a wave. These include diffraction, refraction and colour. By the end of the 1800's the wave properties of light had been worked out in great detail. The aether was widely accepted to be the medium that propagated the light beam, but it could not be demonstrated.

The famous Michelson & Morley experiment.

Few experiments have had such an impact on the future of the world as this one. In 1887 Albert Michelson (1852-1931) and Edward Morley (1838-1923) set up a series of famous experiments where they set out to prove the presence of the aether by measuring the speed of light in the direction of the earth's orbit around the sun, both in the direction that the earth is moving, and at right angles to that direction. They postulated that the earth moves around the sun at such a great speed, that light would appear to be going faster in the direction of the earth's orbit, the speed being the speed of light plus the speed of the earth. At right angles to the earth's orbit the speed would be that of the speed of light alone. The difference should be big enough to be measured with their equipment. They never considered the speed of the sun around the galaxy because galaxies had not yet been discovered. You may well ask how they could possibly do this with relatively crude instruments, by today's standards, because the difference would be very tiny. Well they devised a very clever method using interference patterns. However to their surprise, and to the surprise of everybody else at the time, they found no difference whatsoever. The speed of light was the same in every direction. They repeated the experiment with ever more sophisticated devices but they could never demonstrate any difference. They were exceedingly diligent and they even did the experiment on the top of a high mountain, because they thought the aether might be somehow sticking to the earth. This experiment was later repeated by the German physicist Georg Joos (1884-1954). He also failed to demonstrate a difference in the speed of light. As far as I know, to this day nobody has ever been able to show light moving, in a vacuum, at any other speed, but the speed of light. In transparent materials like glass the situation is very different. In these materials light moves much slower, and this is said to be the reason why a lens bends light. In glass light travels at about 200,000 Km/sec. The ratio of the speed of light in a vacuum to the speed of light in the glass gives the refractive index of the glass. In this case it would be 300,000/200,000 or 1.5.

Albert Einstein (1879-1955), who was then an unknown scientist, addressed this problem. He proposed that the speed of light is a universal constant, and that is the reason why their experiment had failed. He used the experiment of Michelson and Morley as the basis for his theory of relativity. As a schoolboy in my final year, I read Einstein's special theory of relativity. You might well consider this as bizarre behavior today, but times were different, there was no TV and no internet and science and mathematics were regarded as the pinnacle of scholastic achievement. It is a thin little black book and the mathematics is really quite simple. His theory is based on the failure of this famous experiment. I think we would be entitled to call it an infamous experiment because it caused so many problems. It would have been nice to put the negative results down to experimental error but the quality of the work was so good that this was considered to be very unlikely. Einstein postulated that the result obtained had proved his postulate that the maximum speed of light was a universal constant. He gave it the abbreviation c. Ever since that time many experiment have been done to disprove this fundamental tenant of modern physics, but, as far as I know, every experiment to date has failed to prove it wrong. Even today there are many people who believe that it might be possible to somehow go faster than light. Traveling faster than the speed of light is a staple of science fiction. Warp speed has just about become a word in the dictionary.

The speed of light

That the speed of light is a constant is one of the hardest concepts to understand that has ever been published. Let me give an example. If you were in a Second World War fighter traveling at 500 km/hr and fired a gun in front of you at a stationary target on the ground, and the bullet left the barrel at 1000km/hr the bullet would be traveling at 1500 km/hr when it hits the target. It is the speed of the aircraft plus the speed of the bullet. This is simple arithmetic; but this does not apply when you are dealing with light. If you are traveling very fast and shine a light in front of you from a torch, in the direction of your motion, this beam would still travel at a constant speed in front of you and not be the sum of your speed plus the speed of light. In other words it would appear to come out of the torch slowly, but the total speed would be a constant. Whether you shone the light in front of you, or behind you, or in any other direction the speed of the light beam is the same. The speed of light is a constant, and is the rock upon which all modern physics is based. From this postulate Einstein derived the famous equation $e=mc^2$, using relatively simple mathematics. Now this presented a significant problem to Einstein. If the speed of light is a constant, then, in order for other well established equations to work, such as preserving the laws of conservation of momentum and energy, you have to alter other parameters such as time, length and mass.

After applying these 'fixes' he came up with some really bizarre conclusions. If you are in a space craft approaching the speed of light things get really weird. Time slows down for the person traveling in the space craft. People left behind on earth would stay as they are but the traveler would go into the future. Now this sound really farfetched, but can it be demonstrated? Early experiments were done measuring time in moving aircraft which seemed to support Einstein's theory. Today the answer is pretty clear, using GPS satellites. The global satellite positioning system GPS is based on a network of 24 satellites that are orbiting the earth at altitudes of 26,600 Km. It takes light about 0.08 seconds to reach the ground from the satellites. The atomic clocks on board are accurate to about 1 nanosecond per day. The system works by measuring the distance of an object from the satellites. The position is calculated by triangulation. This is indeed a very accurate mechanism for measuring speed, and in has been shown that the speed of light is in fact a constant. The clocks have to be adjusted for the small differences in time as predicted by Einstein's equations. The atomic clocks do 'tick' slower in orbit than on earth. An adjustment is set into the clocks before launch.

Another prediction was that as you go faster your mass increases. This has been demonstrated by accelerating charged subatomic particle to near light speeds. At near light speeds the mass of particles increases. As you pump more and more energy into the particles they just get heavier and heavier but do not increase in velocity. The amount of the increase in mass is what was predicted from the special theory of relativity.

The special theory of relativity deals with the consequences of the speed of light being a constant. On the other hand the general theory of relativity deals with the warping of space-time and gravity. I think that the proven increasing of mass, and time dilation, as you go faster is very strong evidence that the special theory of relativity is correct. Einstein spent many years of his life trying to solve the problem of gravity. In the end I think he came up with a half baked solution of warping of space-time which was wrong. He then covered the error with a thick layer of complicated mathematics which is almost incomprehensible to everyone. In my opinion once you see a paper with a whole lot of Greek symbols, and plenty of calculus you know there is a fundamental problem. I think Winston Churchill was on the money when he said that there are lies, damned lies and mathematics. Well yes, I did substitute mathematics for statistics, but you get the point. When he could not explain the expansion of the universe with his theory, which was only discovered after his theory had been published, he created the cosmological constant. Later he retracted this, and before his death he said the cosmological constant was the biggest mistake of his life. Much later I will address dark energy, dark matter and the expansion of the universe and show that his cosmological constant was pretty close to the truth, and the rest of his theory was actually wrong. This is really ironic. I think he was like a mouse on a wheel in a cage, running faster and faster and unable to get off. A great many praises had been heaped on his head after his general theory of relativity and particle theory of light had been proved. He had been hailed as the greatest scientist that ever lived, after Newton that is, and he was 'expected' to solve the problem of gravity. The pressure was just too great and he caved in. Instead of admitting defeat, he created a fourth dimension, and produced the abomination of warped space-time. However, in my opinion, even though he bombed on his gravity theory, and wrongly dismissed the aether, he still was a very great scientist.

Evolution and the speed of light

Even without any experiments we could have rationalize that the speed of light has a maximum that cannot be exceeded. Our very existence proves it. In our galaxy there are an estimated one hundred billion stars. It now appears that most stars probably do have planets orbiting them. If only a tiny fraction of these have evolved intelligent life there must be millions and millions of planets with intelligent life. Many of these life forms would be far superior in technology to us. We have been technologically savvy for about four thousand years. Imagine how 'clever' we will be in another four million year or four hundred million years. Now if all these advanced civilizations were roaming around with warp speed craft they would surely have landed on earth many times, colonizing it, taking sample, polluting it, and messing up our evolution. Yet, we are here, and the fossil record indicates a clear path of evolution going back at least a billion years. There is no evidence whatsoever, that aliens ever landed on the earth, and there is no evidence that evolution was interfered with, except by natural catastrophes. I cannot conceive of any way that space tourists would not have interfered with our evolution, either by accident or design. The only rational conclusion, I think, is that we evolved despite the best efforts of intelligent aliens to get here, because they could not get to us. Why not? Because they cannot go faster than light, no matter how clever they are, and what technology they have.

We are exceedingly lucky because our sun is located towards the one end, of one of the spirals in our galaxy, where the stars are quite spread out. If we were closer to the center of the galaxy, where most stars are, or if we were in a star cluster, where stars are very close together, then I think we would have had many space tourists visiting us. We would also have been at much greater risk of destruction from super nova explosions and gamma ray bursts, and cosmic radiation. But where we are, is just too far away for any living thing to get to us. Well, you might say this is incorrect because they could travel near the speed of light and live for a very long time. Some trees after all live for thousands of years. Well traveling at the speed of light is just about impossible from an energy point of view. Even if you were able to get to this speed a collision with any object, even a grain of sand would lead to instant destruction.

Let's consider the practical problems of avoiding a collision at a speed close to the speed of light. As mentioned above the speed of light is roughly 300,000 Km/sec. Let's assume that there is a black object the size of a pea in our path. All objects in space seem to be black. I estimate that in order to change course we would need at least 10 seconds to change course, even slightly, to avoid the collision. Anything less than that would cause such enormous forces on the space ship, that it would be torn apart. If we have a ten second margin this means we have to detect the pea at a distance of three million kilometers. Well, you might say we will just use our super powerful radar. Well this won't work. By the time the radar beam gets to the object we will already be on top of it. This is precisely because the speed of light is a constant, and radar travels at the speed of light. Well then, we will just use our massive onboard telescope and look for it. Well this won't work either. Forgetting the fact that the object is black and basically invisible against the black background of space, by the time the light reaches us we will already be on top of the object. Once again this is because the speed of light is a constant. The faster we go the slower the speed of our radar beam or whatever else we try to use. If we had to travel at a more manageable speed of 300 km/sec, which is still considerably faster that the earth orbits the galaxy, a round trip to the nearest star would take ten thousand years. If the pharaoh Ramses had been such an astronaut he would be arriving at his destination just about now, and would be contemplating his equally long return journey. While I am pretty sure our children will see humans landing on Mars, and the moons of Jupiter and Saturn, please don't expect interstellar travel any time soon.

Einstein's conclusion that the speed of light has a constant maximum speed seems to be correct. If it is a constant it means that it cannot be exceeded – period. No matter how unpalatable it might be to have this speed limit imposed on us we have to accept that this is true. If you follow my evolution argument this also means that the human race, and all forms of life on earth, are doomed to extinction. Just as aliens cannot get to us, we cannot get to them.

Why am I making such a fuss about the speed of light in a vacuum? Well the reason is that if it is a constant then there must be a reason for this, and that reason is directly related to the cause of gravity. I will try to show why the speed of light is a constant and why we cannot exceed it.

In a nutshell, Einstein had basically taken the observation of Michelson and Morley that the speed of light was a constant and given this constant a symbol c. In order for other equations to retain their validity, such as the equation for kinetic energy, he derived a few new equations, including the famous energy = mass relationship mentioned earlier. Then, I think Einstein got too big for his boots and went further than it was safe to do. He stated that Michelson and Morley could not demonstrate the existence of the aether because it did not exist. I think this is where he went wrong, and this was the biggest scientific mistake ever made. If the aether did not exist then how was light propagated through space? Well, Einstein said, it was because light was made of particles and particles do not need a medium to move; in fact they would move better without one. Light, he said, was like a stream of very fast bullets. He did not just thumb suck this particle theory of light; he had good solid scientific evidence.

Light as particles

That light was made of particles was not a new idea, and Newton had considered the possibility that light was a stream of particles. He called them 'corpuscles'. He was also aware of the phenomena of interference and had described Newton's rings, which is an interference pattern. So he also knew of the wave properties of light. However Newton never really addressed this problem satisfactorily. Light being made of particles would explain the strange observations of the photoelectric effect. In 1921 Einstein won the Nobel Prize for physics for his work on the photoelectric effect, and not his theory of relativity which many people erroneously believe. Electrons leap off metal surfaces in response to light of high energy, but not low energy. No matter how much low energy light (red light) you add, you cannot shake electrons free. Even a little high-energy light (ultra violet) can have the necessary effect. He postulated that this effect was due to the fact that light was made of particles, and that high-energy particles were needed to kick the electrons of the metal. It also explained the strange observations of black body radiation, whereby a body that is heated will emit light of increasing energy, up to a point, and then the energy of the radiation starts to decrease. The phenomenon of black body radiation was explored by Max Planck (1858-1947) and he explained this phenomenon by light being emitted in packets he called quanta. Einstein postulated the existence of discreet particles he called photons. His statements that light was made of particles, and that the aether did not exist caused a scientific riot at the time. What about all the evidence that light is a wave?

Einstein came along at the right time. Already there were significant problems with the theory that light is a wave traveling through the aether, and people were ready to toss the theory out. Besides the negative findings of Michelson and Morley's experiments, there were other significant problems with the aether theory that were very difficult to explain. Sound travels through air as a compression wave, and this can be easily demonstrated and reproduced in laboratories. The 19th century scientists thought that light travels through the aether in much the same way, namely as a simple compression wave. In a compression wave, molecules move forwards, bumping into other molecule. This causes those molecules to move forwards, and then the original molecules bounce back to where they were. The energy of the wave is thus transmitted by the to and fro motion of the air molecules. A tsunami wave also transmits its enormous power by a compression wave. Water cannot be compressed much, so the wave travels much faster, than a compression wave does in air. In the open ocean during a tsunami there is hardly any movement on the surface of the sea, because the water is being compressed and decompressed in the direction of the movement, and not in an up and down manner. On the surface of the sea however the wave appears to be a transverse wave in that it goes up and down at right angles to the direction of motion. Nevertheless virtually all the force transmitted through water is really transmitted by a compression wave, just as sound is in air.

However there was a problem with light, which meant that it could not be a compression wave. When you pass light through a series of parallel slits, called a diffraction grating, and then pass it through a similar grating where the grating is at right angles to the original grating, the light is completely stopped. As a young trainee pathologist I spent a little time working with a high quality polarizing microscope studying crystals. There is no question that light can be completely stopped with two diffraction gratings at right angles to each other. Figure 1 shows diagrammatically what you should see, and what you actually see, if light was either a compression wave or a stream of particles, when identical diffraction grating are place at right angles, one on top of the other.

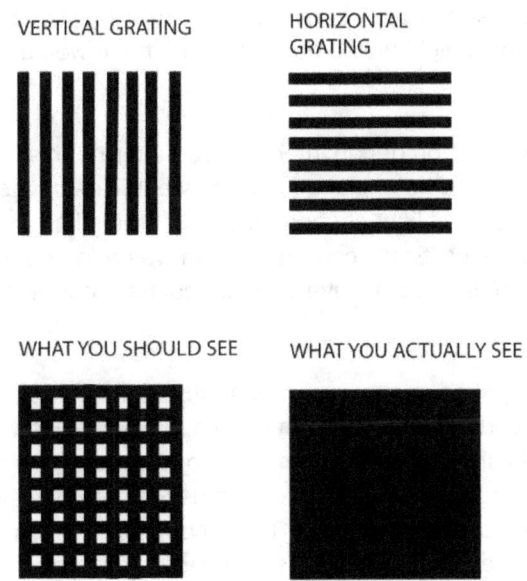

Figure 1   The effect of placing diffraction gratings in front of each other at right angles

If light is a compression wave some of the wave would travel in a straight line through the center of the first slit. This would then also travel in a straight line through the center of the second slit. Some light, admittedly very little, would get through. If it was made of particles then the same logic would apply. Since particles should not bend, then the light would go straight through the centers of both slits. Once again some light should get through. The reality is that nothing gets through. To explain this phenomenon the wave theorists suggested that light travels in an up and down manner as a transverse wave. If light was a transverse wave it would be stopped completely. Light travelling through the first slit would be polarized in the one direction. When this hits the second slit at right angles the polarized light would be stopped completely.

This phenomenon of polarization is now taking on enormous commercial significance with the roll out of three dimensional TV which is based on this very phenomenon of polarized light. This phenomenon of polarization could not be explained easily, and would not occur if light were a compression wave. Polarization knocked a huge hole in the compression wave theory of light, but it also knocked a huge hole in Einstein's particle theory. The same problem confronted those who thought that light was a stream of particles. Some particles traveling in a straight line should go through both diffraction gratings just like light would if it was a compression wave. If it was not a compression wave, but could cause interference patterns, what was it? Unlike sound, light is now said to travel as a transverse wave, meaning it goes up and down as well as forwards. This is very hard to understand. Does this mean that it is constantly changing direction?

Today it is very easy to show that light cannot be a compression wave. Just shine a laser beam at a distant object. If it were a compression wave it would disperse. Like water waves, a compression wave would move out in all directions. Light can only stay parallel over vast distances if it is indeed particles. The distance to the moon has been accurately calculated by shining lasers at reflective dishes left on the moon by lunar astronauts of the Apollo 11, 14 and 15 missions. This would be impossible to do if light was a compression wave. So much light would be dissipated that it would be impossible to detect the reflected light.

If light is made of particles how can you get an interference pattern? This is a good question, and there was no answer then, and there still is no explanation of how particles by themselves could possibly cause an interference pattern. This duality of light stumped the best brains in the world. After years of heated debate a truce was called and the politically correct solution was adopted, that light is both a particle and a wave. All matter is now said to have both wave and particulate properties. It might be good politics but is it correct?

The political correct solution might have convinced most physicists at the time, but it did not convince Einstein, and it has not convinced me, and I am not even a physicist. How can it possibly happen that particles move like a wave? If light is particles then they should move in a straight line according to Newton's laws of motion. Newton's first law of motion states that an object will move at a constant speed in one direction unless an external force acts on it. To be continuously changing their direction would require energy, and that would have to come from somewhere. If it came from the particles they would slow down, which they obviously don't, because light waves can travel for billions of years with hardly any degradation, and as discussed above, the speed is a constant. If light was a wave why would it behave like this, and not go to and fro like a compression wave in water or air?

The Scottish physicist James Clerk Maxwell (1831-1879) postulated in 1861 that the wave behavior of light was due to the action of two different forces acting at right angles to each other; one an electrical force and the other a magnetic force. He called the waves electro-magnetic waves. He produced a series of equations to describe how these two forces could produce a wave. His elegant mathematics kept the peace until Einstein and others produced the particle theory of light. The type of electro-magnetic wave, postulated by Maxwell, was said not to require a medium to propagate it. Now that's easy to say; but once again one has to ask if it is true? Now in Maxwell's time there was no talk of particles of light, and the idea during his lifetime was that light was entirely a wave. But once you introduce the concept of photons, having mass, all be it very small, and being particles, then there are a whole lot of new issues to be addressed and problems to explain.

If you think about this for a bit of time you will come to the conclusion that if photons are moving in a curved path they must be moving through a medium which is causing the curving. An object will require another object in order to change direction. An astronaut stranded, weightless, in space would be unable to turn around unless he could touch something, or could throw something away for him. For a particle of light to continuously change directions it must be coming into contact with something else. If not it would just shoot off into space in one direction and not follow a curved path. There are only two logical explanations for it to be changing directions. The first is that there must be a medium through which the photons are traveling, and pushing against, and the second is that Newton's first law of motion is wrong. In order not to be a victim of cerebral filtering we have to seriously consider the possible that Newton's first law of motion is wrong. Newton derived his laws of motion from the observable universe. He made empirical observations from what he could see and measure and then created equations to explain his observations. When we are dealing with extremely small objects like photons that are traveling at relativistic speed Newton's laws might not apply. A relativistic speed is a speed approaching the speed of light when the strange phenomena of increasing mass, shortening length, and slowing of time appear. At normal speeds which exist on earth none of these phenomena, even though they still exist, are noticeable because they are so tiny. Now I have rather arbitrarily decided to believe that Newton's law does apply, and have therefore chosen to believe that the medium, which was believed to exist for centuries, namely the aether, does in fact exist and that Newton's first law is solid.

When Einstein's particle theory of light was found to be correct, his other statement that the aether did not exists, was also assumed to be correct, and the idea of the aether was shelved. Today most people immediately dismiss the aether theory on the grounds that if it exists, the planets would slow down. This is very simplistic. I think that the mistake Einstein and Michelson and Morley made about the aether was that they did not understand that in order to have an effect, the aether must be accelerating. Understanding aether acceleration is the key to understanding the universe, because all forces are ultimately mediated through accelerating aether.

Ether dynamics

I propose that the aether does exist. It is composed of tiny particles just as others had previously postulated. These particles bounce off each other perfectly and there in no friction.

If the aether has no friction then how can it move a particle? This question lies at the heart of how aether behaves. By friction we mean that things stick together. Aether particles don't stick. When they impact a particle they impart a force during the collision but they don't stick to the particle. There are no other forces of any kind acting between the aether particle and the particle it impacts. After the collision the aether particle moves a little slower and the target impacted particle moves a little faster. The angle and the plane that the aether particle hits the target particle will be exactly the same as the angle and plane that it leaves the target particle. If there is any friction whatsoever then this angle or plane would be different. Well what if it hits the particle at right angles? Well in this case the angle and plane would be the same but this is theoretically impossible. What looks like 90^0 is never exactly 90^0, it may be 89.9 recurring to a power of a hundred but it will never be exactly a right angle.

The magical party balloon

Let's take a practical example of a frictionless environment with perfectly elastic collisions where there is no loss of energy. I place a large round rubber party balloon, which is bigger than your hand, on the floor, and ask you to pick it up and carry it to the room next door, using only one hand. Well this is pretty simple. You put your flat hand on the top of the balloon press down so that it is squashed a little, then dig your fingers into it until you have grip and then pick it up. You then place your hand under the balloon and carry it to the next room. Now I magically remove all friction from the surface of the balloon, and make the rubber special, in that it does not lose any energy when it bounces. I offer you a free beer if you can do the same, namely move it to the room next door. You then try to place your hand on the balloon to squash it a little as you did before, but it immediately shoots away from under your hand. You try to dig your fingers in and get a grip to lift it up, but every time you touch it, it is so slippery that it just moves away. In desperation you try a little cheating and try to hold the balloon with both hands, but that also does not work. Every time you touch it, with one or both hands it just moves away; you just cannot get a grip. After raising quite a sweat, in frustration you kick the balloon, and it flies, like a rocket high into the air, and much higher than you ever thought possible. It hits the ceiling and you are amazed to see that, unlike any balloon you have ever seen, this balloon drops to the ground like a bowling ball, even though it weighs next to nothing. When it hits the ground it bounces high into the air, hits the ceiling and again drops like a bowling ball. You try to grab the balloon but it just slips out of your hands and it is now going even faster. Floor, ceiling, floor, ceiling, wall, floor. I generously offer you a sharp needle and taking hold of the needle you stab the balloon. But to you amazement it just bounces off and the speed increased even more. Floor, ceiling, wall, wall, floor. It does not slow down at all. I suggest that you stab the balloon at right angles with the needle. After several attempts you manage to stab the balloon at right angles and it bursts. Got the bastard! you shout, but to you amazement it does not stop. The deflated balloon carries on through the air, hits the wall and continues bouncing. In frustration you open the window and the last you see of the balloon is it disappearing into your garden being pursued by your now crazed dog. That's not fair you protest, you swindled me, give me another chance. Ok, I relent and change the rubber back to normal rubber, but leave the anti-friction in place. This time you gently flick the balloon into the air and bounce it on the flat of your hand, being very careful to make only small movements. Since the rubber has not been altered the bouncing gradually slows down and you manage, very skillfully, while all the time bouncing the balloon on the flat of your hand, to transport it to the next room where you are rewarded with a well earned beer.

What have you learned about the aether environment from this example? Firstly we have to stop thinking about things behaving as they do in our normal world. Secondly we notice that although there is no friction a force can still be exerted by hitting the object. This example illustrates what I said above. In a frictionless environment there can be no interaction between objects except as a result of an impact collision. Movement can occur by either a particle hitting a particle, or the aether hitting a particle, or a particle hitting the aether. That's it. There are no other possibilities.

How would a body or particle move through the aether? First we have to define what we are talking about. By a body I mean a large object such as a human being or a cannonball or the earth. When I refer to something at the atomic level I use the term particle. Aether acts on individual subatomic particles in an individual manner. The effect of the aether on a body is the sum total of the effects of the aether on the individual sub atomic particles, and vice versa. Now this might seem like splitting hairs but it is not. For instance you may see an accelerating body, such as a spacecraft, and wrongly assume that the particles making up the spacecraft are accelerating because of an accelerating aether flow. As you will see later if a particle is accelerating it must be because of an accelerating aether flow. However the acceleration of the space craft is actually due to innumerable impact collisions at the atomic level. A single impact collision of a single particle would cause the particle to move at a constant speed. Only during the actual impact is the particle accelerating. Thereafter the speed is a constant.

If the velocity of a particle is constant then the aether would be pushed out of the way in front of the particle and the aether would be replace behind the particle. Figure 2 shows this. As the particle moves forwards it pushes aether out of the way but leaves a 'hole' behind it. The aether flows in from behind to fill the hole. Since the aether moves with no energy loss there is not net effect of the aether on the particle.

DIRECTION OF MOTION OF PARTICLE ➤

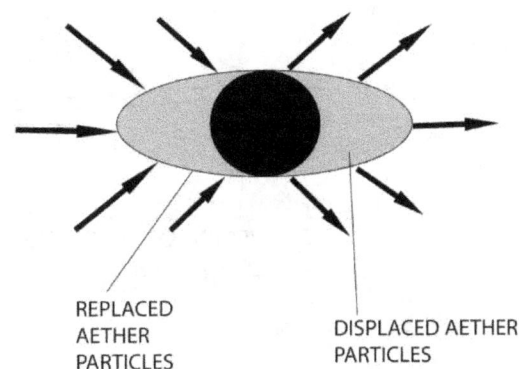

REPLACED
AETHER
PARTICLES

DISPLACED AETHER
PARTICLES

Figure 2   There effect on the aether on a particle moving at constant speed through the aether

The moving particle pushes the aether out of the way in front of it, but is in turn pushed by the aether filling the space left by the moving particle. Since the individual particles of a body moving at constant speed move through the aether with no friction or loss of speed, then the body as a whole also moves through the aether with a constant speed and no loss of energy or momentum.

Now let's look at a large body in relation to Newton's first law of motion. An object will continue to move in one direction, at a constant speed unless an external force acts on the object. Because there is no aether pressure gradient across the particles making up the body, there is no aether pressure gradient across a body traveling at a constant speed. Particles that are accelerating or decelerating are doing so because they are in an accelerating aether flow. No particle can accelerate or decelerate by itself.

Stationary aether will likewise have no effect on a particle or a body. Accelerating aether is however a very different animal. If the aether is accelerating then an aether pressure gradient develops across the individual particles making up the body which causes a force to be exerted on the body. This causes the body to accelerate in the same direction as the accelerating aether flow. The particles and therefore the body will continue to accelerate until the body has the same acceleration, and speed, as the aether flow that it is in. This explains Newton second law of motion which is that force = mass x acceleration. I will come back to this in more detail later. If there is no acceleration then there is no force. When we are dealing with an accelerating aether flow there is a region of relative low aether pressure on the one side of the particle. Figure 3 shows the effect of an accelerating aether flow on a particle.

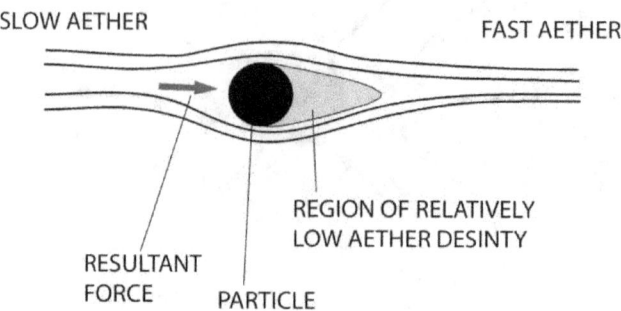

Figure 3   The effect of an accelerating aether flow on a particle

In this diagram there is an accelerating aether flow, caused by aether destruction within a body towards which the particle is being dragged. The aether density is relatively lower on the side of the particle nearer the aether destroying object. Notice that the lines are not parallel. The effect of an accelerating aether flow, like gravity, on a body such as the earth, will be the total of the effects of the accelerating aether flow on all the subatomic particles making up the body.

An aether compression wave surrounds a moving particle

Now there is another important effect of a particle moving through the aether. As it travels forwards it pushes the aether in front and to the sides. This causes a compression wave in the aether acting at right angles to the particle to develop. As the particle passes any given point the aether is pushed back to where it came from, by the surrounding aether, and the pressure is rapidly reduced to what it was. Let's take a stream of photons making up a light beam, travelling at a constant speed as an example. This is shown diagrammatically in figure 4.

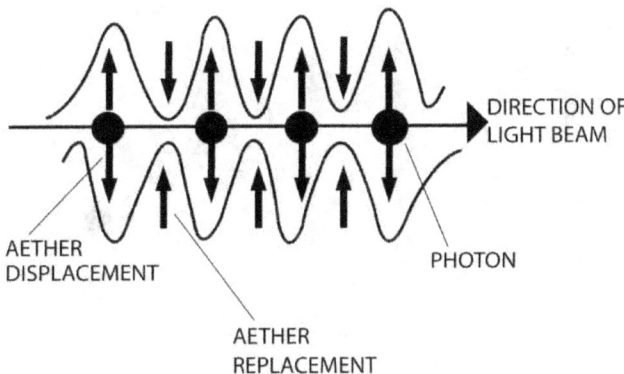

Figure 4   An aether compression wave surrounding a stream of moving photons

As each photon moves through the aether it generates an aether compression wave which moves sideways away from, and then towards the path of the photon. The movement is at right angles to the direction of motion. When a photon moves through the aether the aether is first compressed and then decompressed and there is no loss of momentum or energy in this process.

However, if a photon passes very close to a much larger particle such as a neutron, this compression wave acting at right angles to the direction of the photon will cause an aether pressure gradient to develop across the neutron, and cause it to accelerate at right angles to the direction of the photon. The photon will then have an unequal force acting on it. The wave from the side where there is no object will return to its starting position in the usual manner, but the aether compression wave which accelerated the neutron will not return with the same force. There will be an imbalance in the aether pressure on the sides of the photon and a pressure gradient will develop across the photon from the side, and this will causes the photon to bend towards the neutron. This bending of the photon will occur whenever it passes close to any particle and is the cause of diffraction. The closer the photon passes to the neutron the larger will be the deflection of the photon. It will always deflect toward the nearby object and not away from it. This phenomenon is shown diagrammatically in figure 5.

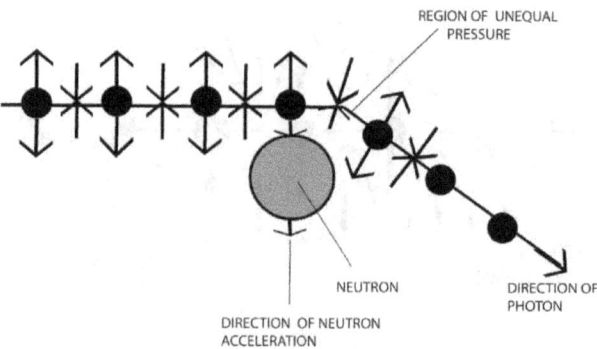

Figure 5   A moving photon passing close to a neutron showing the photon being deflected towards the neutron. Particles are not drawn to scale

The bending of a light beam into the shadow of a straight edge is due to this phenomenon. This causes the generation of interference patterns by particles passing though parallel slits and explains the observed particle wave duality of light. When the direction of motion of photons changes it is because a force is acting on them. This force is due to an aether pressure gradient. Newton's first law is thus observed. It is important to remember that the force causing diffraction of light is due to an aether compression wave. As with any other compression wave the force induced by the wave falls of rapidly, namely by the square of the distance. The effect will only be noticeable at very close proximities. I postulate that this bending of photons around particles, due to the aether compression wave that surrounds a moving photon is not only responsible for diffraction but lies at the heart of other optical phenomena such as refraction, total internal reflection, and the splitting of white light into the colors of the rainbow by a prism. Remember what a photon is. It is a particle with no charge whatsoever. If it has no charge how are you going to be able to alter the path of such a particle through a well defined crystal such as a diamond, where the atoms are well behaved and in essentially fixed positions, except by an aether shove.

What is the speed of the aether compression wave?

The speed of a compression wave depends on the properties of the medium through which it propagates. I am sure that most of the readers have seen movies of hydrogen bomb detonations in Nevada and the Pacific in the 1950's and 1960's. What you see is the bright flash of the bomb followed by the rising mushroom cloud. As the cloud rises you see a circular line on the ground that moves away from the point of explosion. In the case of the explosions in the desert it is clear that this is compression wave of the air. Houses are blown apart as soon as the compressed air wave hits them. This line appears to move at a relatively slow rate. This is because of the large scale. Now it is obvious that the explosion caused objects to move much faster than the compression wave but the speed of the compression wave is constant and is determined by the characteristics of the air and not the atom bomb. The speed depends primarily on the density and compressibility of the air. A moving photons is surrounded by an aether compression wave that moves at right angles to the direction of the photon. This is later followed by a compression wave travelling in the opposite direction so that no energy is lost. The speed of the aether compression wave depends on the density of the aether which is a constant, and the compressibility of the aether which is also a constant. Therefore this compression wave travels at a constant speed. This speed limits the speed of the photon.

Why is the speed of light a constant?

Michelson and Morley proved that the speed of light is a constant. Why should this be so? I propose that the reason why the speed of light is a constant, no matter where you observe it from, is that the density of the aether is a constant, at least in the space between objects. The speed of light is proportional to the density of the aether. Outside the immediate environment of objects the density of aether is a constant. Therefore the speed of light is a constant. Inside objects it is a very different situation. Let's take sound as a comparison. The speed of sound depends on the composition of the propagating medium. The closer the atoms are together the faster the wave moves. If we measure the speed of sound at sea level we come up with a speed of about 730 miles an hour. If you go to the altitude of a jet liner then the speed falls off quite a lot to about 620 miles/hr. Because the density of the aether in the region of the earth, where we are, is a constant, the maximum speed of light is a constant. Notice that I said the speed of light has a constant maximum speed. I did not say that the speed of the aether has a constant maximum speed. You will see the relevance of this when I discuss black holes. Previously I mentioned that if a particle is moving at a constant speed it is because the movement is as a result of an impact collision. If this is the case then the speed of light would be due to this process occurring in the atom, and not the density of the aether. However in the case of photons there must be another factor otherwise time dilation and the other relativistic phenomena would not occur. I think that the speed of light is limited by the speed that the photons can be accelerated within the atom, and this limit is imposed by the density of the aether which limits the speed of the aether compression wave which develops as the photon is accelerated. Although aether can be accelerated and move very rapidly it is not infinite, and there is a limit. This limit is one of the two universal constants. The other is the rate of aether destruction per unit mass which I will discuss later. As particles approach the speed of light the inability of the aether to immediately return to where it was causes the phenomena of time dilation and increasing mass described by Einstein.

How does a photon move in a curving pathway through space?

Light is thought to travel as a wave through space. A photon is universally depicted as a particle moving with a very uniform wave motion in one plane, with a frequency and amplitude. Is this really true?

We can see that because of the aether compression wave surrounding a photon it will bend around another particle when it gets close to that particle, but how can a photon travel as a wave in space where there are no other particles to cause it to bend? This question lies at the heart of the wave particle duality of light. Well I think that the answer is very simple. It can't. When you have thought about this with ruthless logical and have removed all preconceived notions of wave particle duality from your mind you will see that it is indeed impossible. Photons must move in a straight line in space where there are no other particles. If you ask the simple question – do photons have mass you will get a wide range of answers which are devoid of logic and incomprehensible. It is said they have momentum, but no mass, the mass differs whether they are stationary or moving, and they have energy but no mass. There is obviously some confusion over this. According to this theory of gravity, if photons displace aether, which I postulate they do, then by definition, they have mass. (See later). It is extremely small but it is not zero. Photons have mass, and spin but nothing else. If you want to believe that they move, in a vacuum, as a wave, then you have to believe that Newton's first law is wrong. You will now ask that if a photon travels in straight lines in space, then where do the well described wave properties come from. Well I think that the wave properties appear within matter, when they encounter atoms. When photons get near atoms they bend because of the aether compression wave surrounding them. You may say that the photons would then bend according to the type of atom. But this is not necessarily true. Just as all particles fall to earth at the same speed because all matter is made of the same things, (see later) photons bend the same around all atoms because all atoms are made of the same things. Remember that an atom is gigantic in relation to the photon. The wave properties of the photon are displayed when it gets near atoms, any atoms, but the exact characteristics of frequency and amplitude are properties or the photon and not the atom which it is passing. These properties would include mass and surface topographical features. Obviously this is going to cause a lot of controversy and negative comment but I don't want to go further into this as it is off the topic of gravity and it is in any case rather speculative. If light is transmitted as particles, and travels in straight lines, rather than as waves, then the aether does not have to a dense solid crystal as the early pioneers calculated it would have to be. All the prejudice against the aether was therefore unfounded. The speed of light is limited by the speed of the aether compression wave at right angles and not the forwards movement, and has nothing to do with photon wave motion.

The above discussion of light diffraction is a byproduct of the central theme of gravity. They are both due to the presence of the aether.

Can we resurrect the concept of the aether out of the dustbin of history and at the same time explain the failure of Michelson and Morley to demonstrate its existence? Well I think we can. At the end of the paper, after I have discussed planetary motion, I will again turn to the Michelson and Morley experiment and explain why they could not demonstrate a difference in the speed of light.

The nature of matter

You might have noticed that I differentiated between large bodies and small particles. I did this because it is important to think at the subatomic level, because the aether exerts its effects at that level, on an individual particle basis. You might well ask why an accelerating aether flow would affect subatomic particles on an individual basis. The answer can be found n the nature of matter.

At the time all the famous pioneering experiments on light were being done solid matter was believed to be just that - solid. However, in 1889 the New Zealand scientist Ernest Rutherford (1871-1937) showed that this was not so. By bombarding a thin gold sheet with alpha particles, which are basically protons, he showed that most went right through the gold foil, which by itself is surprising, if matter is solid. Even more surprising was the observation that every now and then one would be deflected to the side. Occasionally protons were deflected right back towards the source where they had come from. This showed for the first time that matter is not solid, but that almost the entire mass of an atom is located within the nucleus. Some clever people have developed analogies for this. On a human scale, if one stood in St. Paul's cathedral, then the nucleus of a hydrogen atom would be the size of a walnut, and the electron, the size of a fly, would be whizzing around the walls of the cathedral. The rest would be completely and utterly empty space. Well it would not be empty space it would be pure, unadulterated, aether. Pure aether makes up almost the entire universe.

This 'emptiness' of the atom explains why Rutherford only rarely got a proton colliding with the nucleus.

Bear in mind that a gold nucleus is about as big as you get. This ground braking experiment laid the way to all of modern nuclear physics. Now this gold was matter in its normal state. However matter can exist in different states. It is possible to compress matter much more than it exists on earth. In a neutron star all the matter of the star is compressed into a sphere not much larger than the earth. The entire earth on this scale would be about the size of bowling ball. In a black hole matter might be compressed even more. The reality then is that there is very little matter in space. If matter is mostly empty space, what comprises the universe? I think that almost the entire mass of the universe is made up of aether. Since the distance between atomic nuclei is so vast at an atomic level that when a nearly linear accelerating aether flow bends around an atomic nucleus it comes back to its linear flow within a very short distance after passing the nucleus it is accelerating. In other words because the distance between atomic nuclei is so great the effect of an accelerating aether flow on one nucleus would not alter the effect of the same accelerating aether flow on a neighboring nucleus.

The density of aether

My proposition that the density of aether is a constant begs another question. Why would it be a constant? This is what requires a great leap of faith. I propose that aether particles are created in one or more subatomic particles, and that they are destroyed in one or more subatomic particles. The density of aether within a material object is lower overall than the density of aether in space because aether is destroyed in matter. The balance between the rate of aether creation and the rate of aether destruction will give the resulting density of the aether. If more aether is being created than is being destroyed then the density of aether will rise. If more is being destroyed that is being created then the density will fall. Accelerating aether particles exert force on atoms. The aether particles are very small but there are a lot of them. Aether will move like an ideal gas from an area of high pressure to an area of low pressure. As the aether accelerates it will drag objects along with it. Thus if we look at very large place like the space between galaxies the pressure of the aether will be a constant, because the aether has had a very long time to equilibrate. The density of aether in intergalactic space is higher than in object like the earth and stars because aether is being destroyed within these objects. However most of the universe is space devoid of objects and the density of aether is essentially a constant over vast distances.

Where does the intergalactic aether come from? If the big bang theory is correct then it would have all been created at one time together with the rest of the universe. In this case the possibility exists that there is no more being created and the galaxies are gradually using it up. If the aether was being used up the aether pressure would gradually decrease, and gravity would weaken. In this scenario the universe would expand. Taken to the extreme there would be no aether at all and there would be no gravitational force. If the steady state theory is correct then it would have to be created continuously. In this scenario the universe would be much older and the aether would eventually be completely destroyed, unless it was being replaced. The existence of aether producing particles in space between galaxies could make up for the aether being destroyed in galaxies. As most of the readers know Edward Hubble (1889-1953) showed by measuring the red shift of stars and galaxies that the entire universe is expanding. The further away the galaxy the greater the red shift. When he compared the speed against the distance, measured by a red shift independent method using Cepheid variables, he got a constant, which is now called the Hubble constant. This has stood the test of time. If the big bang theory is correct and the aether is not being replaced then the fall in aether pressure could be causing the expansion of the universe. Nobody has been able to explain the expansion of the universe and the concept of dark energy has been created to explain this phenomenon.

The 19th century scientists viewed aether as a static medium through which light is propagated. I believe that this view was overly simplistic, and that because of this view they missed the true role of the aether. Making light move in a wave like manner, when it is within matter, is just one of the properties of aether. I propose that aether is far more dynamic than this.

Types of sub atomic particles

Matter has been broken down into various subatomic particles. At the time of writing matter appears to be made of six types of quarks, plus neutrinos, electrons, bosons and photons. Each particle has antimatter equivalents. The subatomic particle zoo is almost incomprehensible. I believe that there is difficulty in putting it all together because some particles have not been discovered, and that the dynamics of the fundamental forces, and the role of the aether in particular, have not been understood.

I propose that there can be only three kind of particles namely aether producers, aether destroyers and aether neutral particles. The aether neutral particles do not produce or destroy aether. Every subatomic particle belongs to one of these three types.

The next postulate that I need to make in order to explain the fundamental forces of nature is that aether is continuously made by one group of subatomic particles, and continuously destroyed within atoms by a different group of subatomic particles. Aether produced by subatomic particles just pops up out of nothing and when the aether is destroyed the aether disappears back to nothing. Aether is being continuously produced and destroyed throughout the universe. It is this force of creation that drives the universe. Where this creation and destruction comes from nobody knows but it appears to be absolutely unstoppable.

The aether pressure surrounding an aether producer is higher than the average and decreases as you move further from the aether-producing particle. The aether will move away from the aether producing particle until the density of the aether is the same as that in free intergalactic space. Conversely the aether pressure surrounding an aether destroying particle is lower than the general aether pressure. Aether will move into the lower pressure area until the aether pressures are the same. As the aether moves into the area of low pressure it will drag objects with it. Because of the unequal aether pressures across the object there will be a force on the object which will cause it to accelerate. In a stable system like the solar system I will show that the speed of the aether increases as you get closer to an aether destroying object like the sun, and that this speed follows an inverse square relationship. Conversely in the case of an aether producer the speed of the aether decreases as you get further from the aether producer, also by an inverse square relationship. Later I will demonstrate the relationship between the speed of the aether and the distance to an aether destroyer using the planets, and the moons of Jupiter, and in our own near earth environment. The important thing to realize is that aether moving towards or away from an ether producer or aether destroyer, is that the aether is accelerating. Only accelerating aether is able to exert a force on an object. It is essential to understand this concept of acceleration. Aether moving at a constant speed will not have any effect on an object in that aether flow. This is compatible with Newton's second law, if there is no acceleration there is no force.

The ideal gas laws in relation to the aether.

Now you might want to relate the behavior of aether to the ideal gas law which links the temperature pressure and volume of a gas. I wasted a lot of time myself trying to relate the two but there is one big difference which makes this association not viable. The ideal gas laws work because gasses can be confined and the pressure will rise or fall within the closed container in response to changes in temperature. The keyword is 'closed'. You cannot confine aether. Aether particles are amongst the smallest particles that exist, and as such they can pass around any other particle and cannot be confined. Since they cannot penetrate into the interior of subatomic particles we have to conclude that subatomic particles themselves are made of matter which must be smaller than aether particles. Because you cannot confine aether it will always move from region of high density to regions of low density, and the concept of pressure, at least as it is related to gases is not valid for the aether. Any increase or decrease in aether 'pressure' would be only of a temporary nature as it will always be able to equilibrate. I talk of aether pressure more or less as a synonym for aether density. Any increase in density is of a transient nature and it will equilibrate back to a constant.

The inverse square relationship

One of the main points of Newton's universal law of gravity is the inverse square relationship. It is said that this was first proposed by Robert Hooke (1635-1703). Newton used the idea and never credited Hooke with the concept. If this is true it shows that the behavior of scientists over time is also a constant. Everyone knows the law but nobody has ever given an acceptable explanation of why it exists. The pressure of aether in a given volume is proportional to the number of aether particle in that volume. When an object moves towards an aether destroyer like the earth it travels down a pressure gradient in the shape of a cone. The volume of a cone is given by the equation volume = 1/3 x π x r^2 x h. A volume of a segment of a cone depends on the square of the radius at the base and the square of the radius of the top times the thickness of the segment.

Figure 4 show the concept of the volume of a segment of a cone.

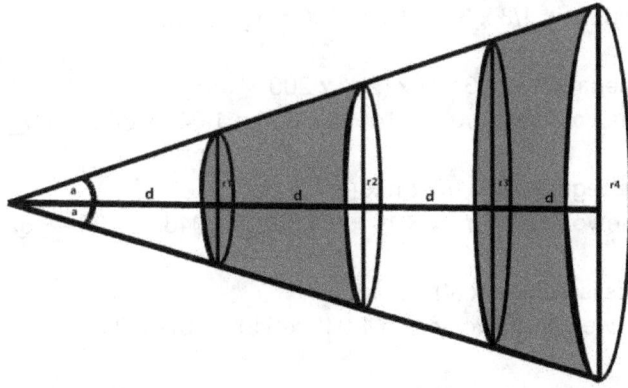

Figure 4   The volumes of segments of a cone in relation to the distance to a point

Let's do a little math on the diagram.
The cosine of the angle a for the first cone = r1 / d.

The cosine for the second cone is r2 / 2d.
Since the angle is the same the cosines are the same the same and r1 / d = r2 / 2d.
Solving this: r2 = (r1 x 2d)/d = 2 x r1.
Therefore the radius r2 is twice that of r1.

Similarly you can show that the radii in the diagram are twice, three times and four times the first radius r1.

Let's get some real figures. Let's make the radius r = 1 cm and d also = 1 cm so that the angle a is 45^0
We will call this the fat cone.

The volume of the first small segment is 1/3 x π x r^2 x d.
Since r = 1 and d = 1 the volume is 1/3 x 3.14 x 1^2 x **1** = 1.047 cm^3

The volume of the first two segment is 1/3 x π x $(2r)^2$ x 2d.
Substituting 1 for r we get the volume = 1/3 x 3.14 x 4 x 2 = 1.047 x **8** = 8.378 cm^3

The volume of the first three segments is 1/3 x π x $(3r)^2$ x 3d
Substituting 1 for r we get the volume = 1/3 x 3.14 x 9 x 3 = 1.048 x **27** = 28.286 cm^3

The volume of all 4 segments is 1/3 x π x $(4r)^2$ x 4d
As before after substitutions we get the volume = 1.048 x **64** = 67.072 cm^3

You immediately see that the numbers in bold is clearly a cubic relationship. Maybe this is due to the shape of the cone.

Let's repeat the exercise using a very long skinny cone. We make d = 100 and r = 1

The volume of the first small segment is 1/3 x π x r^2 x 100.
Since r = 1 and d = 1 the volume is 1/3 x 3.14 x 1^2 x 100 = 104.7 cm^3

The volume of the first two segment is 1/3 x π x $(2r)^2$ x 200.
Substituting 1 for r we get the volume = 1/3 x 3.14 x 4 x 200 = 1.047 x 800 = 837.8 cm^3

The volume of the first three segments is 1/3 x π x $(3r)^2$ x 300
Substituting 1 for r we get the volume = 1/3 x 3.14 x 9 x 300 = 1.048 x 2700 = 2828.6 cm^3

The volume of all 4 segments is 1/3 x π x $(4r)^2$ x 4d
As before after substitutions we get the volume = 1.048 x 6400 = 6707.2 cm^3

We are not interested in the relative volumes of the cubes, what we want are the actual volumes.
Let's take the fat cone first

The volume of the smaller shaded area is 8.378 – 1.047 = 7.33 cm
The volume of the larger shaded area is 67.072 - 28.286 = 38.78
The ratio of the volume of the larger shaded area to the smaller shaded area of the fat cone is thus

38.78 / 7.33 = 5.29 : 1

Let's take the skinny cone
The volume of the smaller shaded area is 837.8 - 104.7 = 733.1
The volume of the larger shaded area is 6707.2 – 2828.6 = 3879.6

The ratio of the volume of the larger shaded area to the smaller shaded area of the fat cone is thus
3879.6/733.1 = 5.29 : 1

We can see that the ratio for the volume of the smaller shaded area to the larger shaded area is the same for the fat and the skinny cones. However you slice and dice the problem the answer is always that the relationship of the volumes is a cubic one. So how come we have an inverse square law, for gravity and not an inverse cube law? The answer lies in what we are actually doing. If we are squeezing a gas from the larger segment into to the smaller segment then the volume of the smaller segment will indeed be related to the volume of the volume of the larger segment in a cubic relationship. But when we are dealing with aether we have an entirely different problem. The density of the aether is already the same in all segments and cannot be compressed further. We cannot talk of compressing the aether. All we can talk about is making the aether go faster. What we really need to do is measure the speed of the aether past a point leaving the density constant.

Let's consider an analogous situation of bulls going into a large pen at the abattoir, to be made into hamburgers. There are four gates into the pen and one is open. You are at the gate and counting the cattle as they go in. Now you need the cattle to go into the pen at a rate of at least 500 an hour or the abattoir will run out of cattle. Now a new train load of cattle arrives and there are hundreds of cattle waiting to go into the pen, but they are going in too slowly. You realize that you can get more cattle through the gate in the same time by either making them walk faster, or you could make them walk closer together and compact them. Initially you could do a bit of both, make them run and make them pack up closer together. You do this and within a short time the bulls are touching each other and there is just a solid mass of bulls going through the gate. At this point further bull compression is not an option, and the only way to get more bulls into the pen given time is to make the bulls run faster or open more gates. This you do. You open one more gate and the number of bulls going into the pen doubles. You open the remaining two and it double again. After that you are left with only one option and that is to make them run faster. You make them run as fast as they can. You see a massive crowd of bulls outside the pen but there is nothing more you can do. How many bulls are outside is now irrelevant all that matters to you is the number passing you in a given time and this depends only on the speed that they can run.

The sun is analogous to the slaughter men inside the abattoir. The sun demands a fixed number of aether particles in a given time to be destroyed. Since the aether is compressed to its maximum the only way you can get more aether into the sun is to make it go faster. As it actually enters the surface of the sun the only thing that matters is the ratio of the number to be destroyed to the surface area of the sun. If this ratio is very high then the speed will be very fast. If the ratio is low then the speed will be lower. Since we are talking about surface areas you can see that this is a square relationship. Let's consider aether moving through a square of the surface of the sun with the length of the sides each being 2 Km. If we want to get that same amount of aether through a square with the length of the side being 1 Km then the aether has to go 4 times faster. There is a square relationship and not a cubic one.

If aether could be compressed like a gas the density of the aether in a segment of the cone would be proportional to the cube of the distance from the point source. But since we cannot compress the aether the only thing that matters is the speed of the aether and this follows a square relationship. This explains Newton's inverse square law. This is an important distinction.

I will propose the basic concepts of the interaction of aether with objects. All the forces of nature whether it is gravity, magnetisms, electrostatic forces, nuclear forces or supernatural forces are fundamentally mediated either through the acceleration of aether or direct particle impact. Nothing, no object, no molecule, no atom and no subatomic particle can move at all unless it is by a direct repulsion due to particle collision, or unless it is in the path of an accelerating aether flow. I will go into these factors in more detail later, but at this stage just read the concepts and keep an open mind. If you ask where I got these concepts from, the answer is I worked them out after thinking about the problem.

1. There are only two types of forces namely repulsive and drag forces. There are three possible types of subatomic particles. Those that produce aether, those that destroy aether, and those that neither produce nor destroy aether. The following rules apply to interactions between aether destroying particles and aether producing particles.

An aether producer will repel every other particle coming close to it except for an aether destroyer. Two aether producers will strongly repel each other as they are both producing aether.

An aether producer will always repel an aether neutral particle.

An aether destroyer will always be strongly dragged towards another aether destroyer.

An aether neutral particle will be dragged towards and aether destroyer.

The interaction of an aether producer and an aether destroyer will depend on the rate of aether production and destruction. If the aether producer is producing aether faster than the aether destroyer is destroying the aether, then the aether producer will repel the aether destroyer. But, by a relatively weak force. If the aether destroyer is destroying aether faster than the aether producer is producing aether then the aether producer will be dragged towards the aether destroyer, also with a relatively weak force.

Two aether neutral particles will not interact directly with each other.

There are no attractive forces of any kind. The electrostatic forces are probably repulsive in the case of repulsion, and almost certainly drag forces due to ether destruction in the cases of apparent attraction. I postulate that quarks are all aether destroyers, electrons are likely to be aether producers and photons and neutrinos are aether neutral.

2. Aether always moves from regions of temporary high aether pressure to regions of temporary low aether pressure in order to keep the aether pressure constant. In this process the aether accelerates towards the aether destroyer. Accelerating aether exerts a force on particles, and in the process can drag much larger bodies with it. Aether never moves, by itself, from regions of low aether pressure to regions of high aether pressure. The increases or decreases in aether pressure as you move to or from aether producer or destroyer is proportional to the square of the distance to or from the aether producer or

28

destroyer. The subatomic particles within any given object are a mixture of aether producers and aether destroyers but in any given piece of matter there are always more aether particles being destroyed than being produced. This excess of aether destruction over aether production prevents the atom from flying apart, and is also responsible for the force of gravity. Gravity and the strong nuclear force are the same thing, only the scale is different.

3. Particles and bodies moving at a constant speed have no effect on the aether, and aether moving at constant speed has no effect on particles or bodies.

Accelerating aether induces acceleration of particles and acceleration of the bodies made up of these particles. The acceleration of the particles is in the same direction as the acceleration of the aether flow.

Accelerating particles cause an accelerating aether flow which can in turn affect other particles. When a particle accelerates, due to an impact collision with another particle, an aether pressure gradient develops across the particle causing a force to develop which acts on the accelerating particle in the opposite direction to the acceleration.

I speculate that close to the surface of aether destroyers and aether producers there are extremely strongly accelerating aether flows, moving towards and away from these particles and these flows form the basis of the strong and weak nuclear forces. Accelerating aether flows, set into motion by accelerating electrons, within magnetic material is the basis of the magnetic force.

4. Aether cannot penetrate into the interior of subatomic particles except to the extent that they are produced or destroyed.

Supernatural forces

Now if light is moving particles, and gravity and the other forces are due to accelerating aether and repulsion by particles, then what of Newton's idea that light was a compression wave in the aether. I think that by now we realize that Newton was no dummy. He thought that aether has many properties similar to a fluid or a gas. Like a fluids or a gas there should be an aether compression wave. Well I agree. If aether is essentially tightly packed incompressible particles then it should be able to transmit a force by a compression wave. There are then two problems. How can you actually produce a compression wave, and how would you be able to measure it. None of the forces mentioned above are due to the propagation of an aether compression wave.

To produce a compression wave you have to remember the central tenant to this theory, that an object is set into motion by an accelerating aether flow or an impact collision. A stationary object or an object traveling at a constant speed has no effect on the aether. To get a compression wave, or a series of compression waves the object would have to accelerate and then decelerate very rapidly by a non aether flow mechanism. During a particle collision a particle will go from rest to a high speed during the actual collision. This could cause an aether compression wave. An electron changing orbits could be such an accelerating object. But according to modern theory a photon is emitted when this happens. Possibly a compression wave is also produced. Electrons jumping across the gap of a spark plug are also objects rapidly accelerating and decelerating. In addition to emitting photons they could probably also cause an aether compression wave. When a nuclear reaction takes place and protons fuse together to form new atoms it is possible that the subatomic particles involved in these reactions would be accelerating and decelerating rapidly and could also produce aether compression waves. It is thus likely that aether compression waves do exist but have not yet been detected. As I mentioned earlier a moving photon, or for that matter any particle, is surrounded by a to and fro aether compression wave which moves at right angles to the direction of the photon.

Detecting an aether compression wave presents an entirely different problem. The instruments that we have built all detect different properties of the aether. A chemical balance detects the stable slowly accelerating aether flow that is gravity. A compass detects the accelerating aether flows induced in magnetic materials by accelerating electrons. Photographic plates detect the impact of photons. We have no instruments to detect aether compression waves. The only thing we have that can detect one at all is an interference pattern. Just because we cannot measure them does not mean they don't exist. It is possible that animals have evolved some mechanisms to detect aether compression waves. Let's take an example. Prior to an earthquake is had been frequently reported that animals exhibit strange behavior. Fish jump out of water, horses and dogs leave their homes and snakes leave their holes, even in the middle of winter. Nobody has developed a device capable of picking up the phenomena responsible for this behavior. It is not beyond the realms of possibility that when rocks are subjected to the enormous forces involved in an earthquake that rapid rearrangement of molecules could occur in the rocks. We know that in an asteroid impacts the structure of quarts changes at the site of the impact into what is known as shocked quartz. A similar molecular rearrangement could occur before an earthquake. The piezoelectric effect shows that electrons can be released from rocks under pressure. It is very unlikely, as has been suggested, that animals are picking up minute electric currents in the ground, as rock is a very good insulator. Furthermore nobody has been able to detect any electrical currents in these situations. However, it is possible that aether compression waves are being generated due to rapid re-arrangement of atoms within the rock and that these aether compression waves are being detected by animals, in some unknown manner. Since it is a compression wave we are talking about, the range where they could be detected would be fairly short. An aether compression wave would radiate out in all directions, and the strength would fall off with the square of the distance from the origin. The effect would only be detected very close to the site of the quake and not hundreds of miles away.

By changing the frequency of the compression waves it would be possible to transmit information in a similar manner to the way we transmit information by electro-magnetic weaves. An aether compression wave would also be able to transmit a force and be able to move objects. The compression wave would be able to move through 'solid' objects. I don't want to get too bogged down in a discussion of supernatural phenomena because it is off the topic of gravity and I don't know much about it. I present it here as a possible rational way to explain the mechanism of these phenomena.

Galileo and his tower

I shall now turn to the most important gravity experiment ever undertaken which leads directly to many of the important concepts in this theory of gravity. It was performed by the Italian renaissance scientist Galileo. I shall demonstrate how these famous experiments have direct relevance to the theory that gravity is due to a slow accelerating aether flow.

In a series of famous experiments in 1633 Galileo showed that all objects fall to earth at the same speed regardless of what they are made of. Although there is some controversy about where or how the experiments were done it is probably true the Galileo performed the experiments from the leaning tower of Pisa. Of course it was leaning a lot less 370 year ago than it is today. He found that it does not matter what the object is made of, or how large or small it is, or what shape it is, all objects fall at the same speed. I believe that even today, most of the general public would not believe you if you tell them. It just does not seem to make any sense, that in a vacuum a feather and a cannon ball will fall at the same speed. In a famous picture the Apollo astronauts Dave Scott and Jim Irwin dropped a feather and a heavy prospector's pick on the moon. As predicted they hit the ground at the same time. This is very counter intuitive and for centuries this was a puzzle. I thought about this myself long and hard and finally I worked it out for myself. This question still causes considerably confusion. If you do an internet search on this question you will see many different answers, some of which are plainly wrong, and many answers are very vague. The key to understanding this is to realize that gravity does not act on an object, it acts on the subatomic particles that make up the object. Once you know that all materials are made of the same things, namely protons and neutrons, and that these particles in turn are made up of even smaller particles the answer is obvious. The gravitational force is not acting on the cannon balls, or the feathers, or the prospectors pick, or whatever else you choose to drop; it is acting on the subatomic particles that make up those objects. Since the subatomic particles in all matter are the same, they all fall to earth at the same speed. What matters is the force per subatomic particle. Since this is the same the acceleration per particle is the same, and the final velocity is the same.

According to the big bang theory all particles were made at the same time. It is quite a thought that the subatomic particles that make up your body, are exactly the same as the subatomic particles that make up everything else, and that they are all the same age, about 14.5 billion year old.

Now how does this fit in with my aether theory of gravity. Well, aether is being destroyed by subatomic particles that are all the same. The rate of destruction by aether destroying particles is a constant for that type of particle. The aether flow into all atoms is therefore the same, because it is flowing into the subatomic particles. Therefore the force of gravity will have the same effect on everything regardless of its composition or shape. You are not attracted to the earth by gravity; you are pushed onto it by the aether that flows through you during its headlong rush into the center of the earth. Sure the atoms in your body will destroy some aether, but this is very small compared to what the earth is destroying. The larger the object the more aether will be destroyed by the object in a given time period. The greater the aether destruction the greater will be the acceleration of aether into the object and the greater the drag on any objects in its way.

If you are careless in your thinking, you could easily conclude, as I initially did, that the force exerted by this flow will be proportional to the speed of the aether, and the density of the aether. But, this is wrong. The speed of the aether is completely irrelevant, what matters is the acceleration of the aether. This is the mistake that Einstein made with his interpretation of the Michelson and Morley experiment. The acceleration, and not the speed of the aether flow, will be proportional to the difference in aether pressure. Close to the surface of the actual subatomic aether destroying particles the acceleration will be high and the drag per unit area very great. At a great distance the acceleration will be low and almost un-measurable.

It does not matter how hot the object is, or how much radiation is pouring out of it, or what the object is made of; aether will relentlessly accelerate into it, dragging other things along with it. The sun, despite its production of hot gases and charged particles and its heat, is a net destroyer of aether particles. Aether is pushed towards the sun from the surrounding space to replace those particles that are destroyed. If our planet were not moving very fast, in orbit around the sun we, and the whole earth, would be dragged into the sun by the aether flow as it accelerates into the sun. The flow of aether into all atoms, and its destructions within those atoms is the same for all atoms. Hence all objects fall at the same speed.

There are some more fundamental conclusions that can be made from Galileo's experiment. When an object falls it does not break apart, or change colour or become charged. This means that the forces holding the subatomic particles together, as well as the forces holding the molecules together, must be much greater that the force of gravity. Even when stars are subjected to extreme acceleration as has recently been demonstrated by giant stars close to the center of our galaxy these stars do not change in appearance. However this is not to say that an accelerating aether flow cannot break up an object that is loosely held together. This was shown by the gravitation breakup of the comet Shoemaker-Levy 9 when it hit Jupiter in 1994. The comet was subjected to two accelerating aether flows at the same time. One going into the sun, and one going into Jupiter. The forces moving in opposite directions became too large for the comet to bear and it fragmented into numerous smaller pieces.

Mass

Now that you have an idea of what gravity is I will tackle the problems of mass, momentum and energy. As with gravity very good, strong and robust equations built on observations, have been made to predict the behavior of matter. Once again it was our friend Newton who created these equations. Unfortunately once again he was fundamentally wrong, and very vague about what is actually going on.

In order to understand mass, momentum and energy you have to accept the postulate that aether cannot enter into the interior of subatomic particles except, to the extent that they are created or destroyed. If they did then my theory would not work. I am a medical man by training and I know that life on earth depends on the cell membrane which keeps the inside of the cell separated for the outside of the cell. The concentration of the chemicals in our blood is kept constant by the kidney and other mechanisms so that it remains close to what it is in seawater. As life evolved the cell membrane separated the inside of the cells from the sea water outside. And it still does the same today. Except that our cells are now surrounded by blood, and not sea water. But the electrolyte concentration in blood is very similar to sea water. Because the cell membrane keeps certain ions such as sodium outside the cell, and other ions such as potassium inside the cell, an electrical potential develops across the cell membrane, and this is responsible for things like nerve conduction. The cell requires energy to keep certain ions out of the cell and others inside. When a cell dies the membrane becomes leaky and ions flow freely across the membrane. The electrical potential disappears, water and ions rush in and the cell dies.

I believe that the surface of subatomic particles does much the same thing. It keeps the aether out and other goodies inside. Now you might say that it is a wild speculation to say that subatomic particles could have definite physical structures like cell membranes do. But why not? To have a structured and orderly universe, surely you must have structured and orderly building blocks. Be that as it may if aether could move freely in and out of subatomic particles none of the forces I have discussed would exist, because an aether pressure gradient could not develop across a particle.

Let's assume that we have a rocket ship loaded with fuel ready to go to Jupiter. The ship is at rest in relation to the earth and to everything else surrounding it such as the loading docks and astronauts. Now if we want to get to a speed of say 100,000 km/hr how much fuel do we need to burn? Well this is simple; we just look it up in empirically derived tables using the mass and the terminal velocity. But how would you derive this from first principles. Well you could work out the kinetic energy that it will attain according to the formula $E = 1/2mv^2$ where e is the energy, an m is the mass and v is the final velocity. But what exactly are these things.

Let's go into this in a little more detail relating this to the aether. A fundamental tenant of the theory is that aether cannot penetrate into the interior of the subatomic particles making up an atom. These particles may be making aether, or they could be destroying aether, or they may be aether neutral, but the number of particles being produced or destroyed by the atom as a whole is small in relation to the size of the atom. In order for the model to work the relationship between the sizes of an aether particle to a subatomic particle would be about the same as the relationship of an air molecule to us. The spacecraft would be displacing aether particles because aether particles cannot penetrate into the interior of the subatomic particles making up the spacecraft. The greater the number of subatomic particles the greater the number of aether particles that will be displaced. When the ship accelerates the number of aether particles that would have to be accelerated and pushed out of the way would be this number.

Archimedes worked out thousands of year ago that a floating ship displaces its own weight of water. A 90,000 ton tanker displaces 90,000 tones of water. In an analogous manner we could say that the 'mass' of an object is equal to the 'mass' of the aether particles that it displaces, but this does not help since we are trying to define what mass is. Since light behaves exactly the same wherever it is it, it is reasonable to assume that aether particles are all the same. If aether particles are all the same then we can say that the mass of an object is equal to the number of aether particles that it displaces. Since the density of aether is a constant then we can also say that the mass of an object is also equal to the volume of the aether that it displaces. Obviously the subatomic particles themselves are made of something but there is no way of knowing what that is. All we can say is that it must be smaller than aether. Now, if a subatomic particle displaces say a hundred trillion aether particles when in it is at rest it will displace the same number when it is moving. Now we know that the mass of an atom is located mainly in the nucleus and that this is very small in relation to the size of the atom. Almost the entire space occupied by an atom is aether. From this analogy we can see that almost the entire mass of the universe is made up of aether. The volume of the aether in an object like a human is so large compared to the volume displaced by the subatomic particles that comprise a human that on a comparative basis the particles that make up a human are almost meaningless in comparison. When aether starts to accelerate, even by a tiny amount, the momentum of the moving aether is so great that it is almost unstoppable, and moving stars and planets around is a walk in the park.

Now let's see what happens when we start to move our space ship to go to Jupiter. The first problem is how do you get it to move. Now remember to think in terms of subatomic particles and not the ship as a whole. Getting something to move in our world is so easy that you don't even think about it. When we want to walk we simply put our foot forwards and walk. But in order to move forwards we are actually pushing the world backwards. We cannot see our effect on the earth because it is so big in relation to ourselves. But if you look at an artillery piece firing a shell you can see the effect. As the shell moves forward the gun recoils backwards. Notice that the recoil only occurs while the shell is in the barrel and accelerating. As soon as the shell leaves the barrel and stops accelerating the recoil stops. This is Newton's third law of motion in action. This states that for every action there is an equal and opposite reaction. But why is there an equal and opposite reaction? Newton conveniently forgot to tell us why. A subatomic particle in the aether is in the same situation as an astronaut stranded in space. In order to turn around, or move, he has to be able to touch something or throw something away from himself. If we examine this situation in relation to the aether then we can see why this happens. In order for a subatomic particle to accelerate we have to be able to accelerate the particle with an accelerating aether flow or hit the particle with another particle.

On a large scale what we observe is the spaceship accelerating in one direction and the hot gases accelerating in the opposite direction. Once the fuel is used up the acceleration will stop and the spaceship will continue moving at its final constant speed. Let's consider what happened when our spaceship fires its rocket motor. Hot gasses shoot out of the back of the rocket and the rocket accelerates forwards. These fast molecules of gas appear to be moving into empty space but they are not moving into nothing, they are moving into the aether. If we were to measure the speed of the gas at a fixed point in space behind the rocket we would find that it is not accelerating. Although the rocket ship is accelerating the individual gas molecules are not accelerating. After it has left the rocket motor the gas is traveling at a constant speed. We can also be pretty sure that the force exerted by the gases on the rocket occurred during impact collisions, and that the rocket is not moving because of an accelerating aether flow. At the time of the impact the gas atoms were accelerating away from the rocket. This acceleration occurred only during the actual impact. They were not accelerating into nothing they were accelerating into the aether. This caused the aether to compress. This resistance causes the gas molecules to pile up and exert a force on the rocket, which pushed the rocket forwards. If there was no aether to offer resistance the gas molecules would just zoom off into space and there would be no force on the rocket. The presence of the aether is responsible for the resistance to acceleration which is responsible for Newton third law of motion. For every action there is an equal and opposite reaction. You may well ask what were the particle impacts that got the gas atoms moving in the first place. Well I have no idea, but I suspect that this phenomenon must lie at the heart of all exothermic chemical reactions.

Let's consider momentum in relation to the aether. The momentum of an object is the mass of an object multiplied by the velocity. The mass of an object is the number of aether particles that the object displaces. Therefore the momentum of an object is equal to the number of displaced aether particles multiplied by their average speed. When the rocket fires its motors we perceive that the fuel is used to accelerate the craft. The fuel is actually used to push apart the aether in front of every single subatomic particles that the ship if made of. At the same time it is pushing and accelerating the aether in the opposite direction.

The next time you are in an aircraft as it takes off think of this. As the engines go to full throttle and the plane accelerates down the runway you are pushed into your seat and as it takes off your blood seems to go into your shoes. What's pushing you? There is no airflow in the cabin, so it's not air resistance. You are being pushed by the resistance of trillion of trillions of aether particles that stubbornly don't want to move out of the way to let your accelerating atoms pass.

In the same way when you accelerate your car you are using petrol to move air molecules out of the way to overcome air resistance, and you are also moving aether particles out of the way to overcome aether resistance. If you make your car out of aluminum instead of steel the air resistance will not change but the aether resistance will be much less and you will save fuel. Aluminum has fewer subatomic particles than steel for the same volume and will displace less aether. When your car is traveling at a constant speed on the highway your fuel consumption is related only to the air resistance and tire resistance.

Because air and tire molecules interact with each other you must constantly add more fuel to keep pushing these away. However the composition of the vehicle, aluminum or steel makes no difference because you do not need to keep adding energy to push more aether out of the way. Aether particle collisions are frictionless and there is no loss of energy at a constant speed. Your car behaves just as a single photon would as it travels through the aether. To really understand mass you have to stop thinking about your car as a car and start thinking about it as a massive numbers of individual tiny subatomic particles. The behavior of the car as a whole depends on the behavior of the individual particles. Each one pushes the aether out of its way. The aether then flows around it and comes in from behind. Since there is no friction each particle does not slow down and the car does not slow down. The momentum of the aether particles depends on the number of particles being displaced and the velocity of these particles. If you want to go faster you have to increase the momentum of the particles being pushed. This requires extra energy.

Energy

Energy is another concept that has been bandied about for centuries but nobody really knows what it is. How I see it is that energy is similar to momentum, but whereas momentum is the product of the speed times the number of aether particles displaced, energy is related to the generation of an aether pressure gradient. Without an aether pressure gradient across a body the body will not accelerate. The greater the pressure gradient, the greater the acceleration. But it is not only the magnitude of the acceleration that matters it is also the duration of the acceleration. I view energy as the ability to create momentum, because once the acceleration stops, what we are left with, is a body with a new constant speed and a new constant momentum.

The more aether particles that are accelerated the greater the energy expended. The greater the aether pressure gradient the faster the acceleration and the greater the energy expended. The longer the pressure gradient is maintained the greater is final velocity and the greater is the energy expended. The number of aether particles being displaced by the object is known as the mass. The acceleration of the object depends on the aether pressure differential across the object. The energy expended in accelerating a body would then be given by the equation $E = n \times p \times t$ where E is the energy expended in moving a body, n is the number of aether particles displaced by the body, p is the aether pressure gradient across the object and t is the time that the aether pressure gradient is maintained.

Now let's see if we can use this equation to derive the formula for gravitational potential energy that we are all familiar with. In our equation n is the number of aether particles that are displaced by the accelerating object. As mentioned previously this is known as the mass of the object. p is the aether pressure gradient across the object. The acceleration of the object is directly proportional to the aether pressure gradient. We can therefore substitute an a for the p, a being the acceleration. The distance the object travels is proportional to the time that it is accelerating. We can therefore substitute a symbol d for the time. The equation can therefore be rewritten as $E = m \times a \times d$. In the case of gravitational potential energy the acceleration is given the symbol g and the distance by h being the height. The gravitational potential energy of an object is thus given by the equation $E = m \times g \times h$ where g is the acceleration of the object and h is the height above the ground. We have thus derived the standard formula for gravitational potential energy from the basic laws of aether motion.

Volume surface area relationship

Since aether is the result of a drag force would the surface area of the particles being accelerated be of importance? Well I think that it would be. If the surface area is large, then it would take longer for the aether to flow around the object. Let's take the analogy of an airplane wing. When the air goes over the top of the wing it takes longer to travel than when it is going underneath the wing, because the wing is curved on the top. Because of the difference in air speed there is a resultant upwards force on the wing and this gives the wing lift. A similar thing would probably happen with aether. When we think of subatomic particles, say protons and neutrons we automatically think of them as being small spheres. This is because everything we see in space is either spherical or made of spheres. But there is no evidence at all that this is true for subatomic particles. Now you might take issue with that saying that photographs of gold atoms have been taken with scanning elector microscopes that clearly show round atoms. But what they are seeing is metal deposited on the surface of the atom. They are not seeing the nucleus. But it is reasonable to say from these pictures that the entire atom is round. Subatomic particles could be very plastic and take on any shape that is convenient for them. A single subatomic particle is probably spherical because it is surrounded by aether pushing on to it from all sides. If the assumption that there is no aether inside a subatomic particle is true then, there would be an aether pressure on the particle causing it to become spherical. However if you combine two subatomic particles together then they are unlikely to exist as two spheres stuck on to each other. A more likely scenario is that they would exist as two hemisphere stuck to each other so that the combined surface area is at a minimum. In this view subatomic particles are plastic in nature and can take on any shape that gives the minimum surface area for the combination. Three subatomic particles stuck together could exist as two hemispheres and the third smeared out over the surface of the other two. Four could exist as four equal segments of a sphere. There are many ways that they could combine, but only one which would actually give the lowest surface area.

I will give a simple example. Let's take two spherical bodies combining to form a new body made up of two hemispheres fused together. Let's do some math. The surface area of a sphere is given by the formula $A = 4 \times \pi \times r^2$ and the volume of a sphere is $4/3 \times \pi \times r^3$. Let's start with two spheres each with a radius of one cm. The volume of each sphere is therefore $4/3 \times 3.14 \times 1^3$. The volume of each sphere works out to be 4.19 cm^3. The surface area of each sphere is $4 \times \pi \times 1 \times 1 = 12.56$ cm^3 and the surface area of both is 25.13 cm^2. Combine these two into a new sphere. The volume of the new sphere is the sum of the two spheres which is 8.38 cm^3. We know the volume of the new sphere and we can calculate the new radius. This works out to be 1.26 cm. From this we can calculate the new surface area. If we use the new radius of the fused spheres to calculate the new surface area we get $4 \times \pi \times 1.26^2 = 19.95$ cm^2. Rounding these figures we see that the area drops from about 25 cm^2 to 20 cm^2. A drop of about 20%.

The detailed figures are given in table 1.

Initial volumes of 2 spheres	8.3776 cm^3
Initial combined surface area of both spheres	25.1328 cm^2
New radius of fused spheres	1.2599 cm
Volume of fused spheres	8.3776 cm^3
New surface area of fused spheres	19.9472 cm^2

Table 1   The effect on the surface area of fusing two spheres each with a radius of 1 cm into a single

new sphere

As you can see when two particles combine to form a new particle with the same volume there is a significant decrease in the combined surface area. This would have a big effect on particles in a strong accelerating aether flow. In the nucleus of an atom, four particles which were stuck together because of an accelerating aether flow, might now fly apart if reduced to two particles.

The heavy water paradox

Now let's make a rather unusual diversion and consider heavy water for a few minutes. Trust me there is method in my madness. It was discovered in Norway that when you electrolyze water to split it into hydrogen and oxygen that after a long time there accumulates a residual amount of water that cannot be spilt easily into hydrogen and oxygen. This water appears to be normal water but when this water is frozen into an ice cube and placed in a glass of water it sinks to the bottom of the glass and does not float like regular water. It was later shown that this heavy water was made of oxygen and deuterium and not oxygen and hydrogen like regular water. Deuterium is an isotope of hydrogen and consists of one proton and one neutron. The molecular weight is higher and it is heavier. Normal ice floats when it is frozen because the atoms move further away from each other, and the density decreases, and it floats. When heavy water if frozen it also expands on freezing but the expansion is not enough to make up for the extra weight of the extra neutrons and it sinks in water.

Let's build a Galileo type tower, ignoring air resistance, and see what happens when you drop regular ice, and water from the tower. Well you say they will hit the ground at the same time because Galileo showed that five hundred years ago, and it is because everything is made of the same protons and neutrons. Gravity acts on the subatomic particles, and since they are all the same all objects will accelerate at the same rate. Right? Yes, that is right and when I drop them for the tower I find that normal ice and normal water hit the ground at the same time. Now what will happen when I drop heavy water and ice made from heavy water? They will hit the ground at the same time. Right again. What will happen when I drop a cube of regular ice and a cube of ice made from heavy water? They will hit the ground at the same time. Wrong, don't be so thick. But you said that everything accelerates at the same rate because they are made of the same things? That is right but heavy water is not made of the same things as regular water.

There is an extra neutron in deuterium and a neutron is not a proton. Now a proton carries a positive charges and a neutron does not. This shows that there must be at least one fundamental particle that is different between a proton and a neutron. But, you say a neutron is probably just a proton that has captured an electron. If you add the masses of the two together then the combined mass of the proton and the electron is about the same as the neutron and when you add the positive charge of the proton to the negative charge on the electron then you get a neutral charge, the same as the neutron. Well this might or might not be true, but from a gravity perspective it does not make a difference. They are still different. Remember that gravity is a drag force and this depends on the surface area of the particles in the accelerating aether stream. The smaller the particle the greater the relative surface area and the greater the relative drag force exerted on the particle. The sum of the drag forces on a separate electron and a separate proton will be greater than the drag force on the combination of a proton and an electron that have combined together into one larger particle. Because the drag force is less on the neutron than on the combined drag forces on an electron and a proton the neutron will accelerate slower. It is like a sailing ship in the wind. A folded up sail will have the same mass as a sail that is spread out, but it will not catch the wind. Ice made from normal water should hit the ground first. The difference would be very tiny but could be measured with atomic clocks.

But, you may say the heavy water is heavier, much heavier than normal water. You can measure this easily with a chemical balance. Don't be fooled. With a chemical balance you are measuring the number of subatomic particles within a fixed volume. Sure one cubic centimeter of heavy water will be heavier that one cubic centimeter of normal water because there are a lot more subatomic particles. But when an object falls under the influence of gravity the number of subatomic particles per unit volume is not important. Rather the composition of the particles is important. This is what Galileo proved 500 year ago. In normal matter the composition of all matter is basically the same. The ratio of protons to neutrons, and therefore the ratio of quarks etc. which make up the protons and neutrons are also the same in all matter. With heavy water we are not dealing with normal matter. Abnormal matter has been deliberately, and with great effort accumulated. When uranium 235 is separated from uranium 238 to make an atomic bomb a similar enormous effort is expended to get hold of the abnormal matter. This abnormal matter will behave differently than normal matter in an accelerating aether flow.

You cannot demonstrate the difference in a centrifuge. Although manufacturers will tell you that their machine can generate 60,000 times the force of gravity it is doing no such thing. There are no aether flows in a centrifuge, just a density gradient due to centrifugal force. The acceleration due to gravity on the surface of the earth is roughly 10 m/sec^2. To get to a true 60,000 times gravity you would need an accelerating aether flow that is accelerating at the rate of 600,000 m /sec2. If this acceleration was maintained for just 8 minutes the aether would be moving at the speed of light, which is about 300000 Km/sec. At the risk of sounding boring gravity is due to the acceleration of the aether, and not the speed of the aether. What then is happening in the centrifuge? Well the force holding the head of the centrifuge together, which allows the density gradient to develop, is not coming from gravity, it is coming from the intermolecular forces holding the steel together, and this has to be strong, really strong.

The bending of light by massive objects

Now let's consider the bending of light. Many years ago when I was visiting Basel on the Rhine I was crossing the river on the bridge. There was a small crowd gathered at the side and I stopped to see what was going on. There was a swimmer who was swimming across the river. Now if you have ever seen the Rhine you will know that this is something you do not see every day. Every now and then he would stop to find where he was by looking at the far bank and then he would adjust his direction by swimming against the current. He probably spent more time and energy fighting against the current than swimming across the river. Now in Michelson and Morley's experiment they regarded light in the same position as the swimmer crossing the Rhine. The photons are traveling through the aether in one direction but the aether is moving at right angles to it. The aether flow, they reasoned, should cause the path of the light to be deflected, and the deflection would prove the existence of the aether. But by now you know the error. The swimmer crossing the Rhine was deflected from his course by the force of the water. The water was not accelerating, but he was nevertheless deflected because, when water hits a swimmer the encounter is not inelastic. The water does not go through the swimmer and hit the subatomic particles on an individual basis as aether does. Water molecules interact quite strongly with each other and with skin. If you look at the meniscus of water in glass the interaction is clearly seen. The analogy of the swimmer in the water is quite inappropriate for the deflection of a light beam. To deflect the light beam the aether would have to be accelerating, and accelerating very rapidly at that. The accelerating aether would cause the bending of the light by exerting a force on each individual photon in the light beam.

When a light beam passes a large object, such as the sun, light is bent by the object. Einstein predicted in his theory of relativity that light would be bent by a massive object like our sun. His prediction was confirmed during a total eclipse of the sun. Comparing two photographs of stars in the region of the sun a measurable shift in the positions of the stars was observed. Now that we know that aether particles are accelerating into the sun at high speed the bending of light seems pretty obvious, because photons, like any other subatomic particles, have mass in that they displace aether, and will accelerate in an accelerating aether stream.

Let's try to work out the bending of light as it passes the earth using what we now know about gravity. Now as I promised you in the beginning I am not going to use any heavy mathematics, just a rough calculation is all that is needed. The speed of light is about 300,000 km per second. That means that the photons are traveling through the aether at that speed. When a beam of light passes close to the earth it would be influenced by the aether streaming into the earth for about 0.1 sec. During this time the beam would have traveled 30,000 Km which is more than the diameter of the earth. The acceleration 'due to gravity' is a well known figure, namely 9.8 m/sec^2. Now we know from the above discussion that the speed of the aether particles are not important, what matters is the acceleration. Since all objects fall to the earth at the same speed, and photons are objects, then light also falls at the same speed as a cannon ball. The distance an accelerating object moves in a given time is given by the equation $s = ut + \frac{1}{2} at^2$ where s is the distance, ut is the initial speed t is the time and a is the acceleration. The initial downward speed of light is zero therefore in 0.1 sec the distance that light falls is 4.9 x 0.1 x 0.1 = 0.049m or about 5 centimeters. Clearly over a distance of 30,000km a fall of 5 cm, while not being nothing is probably immeasurable.

Consider the situation with light traveling past the surface of the sun. The sun has a diameter of 1,391,000 km. A light beam would take just over 4 seconds to pass, at right angles, across the surface of the sun. Now the acceleration due to gravity of the sun is much greater than the earth. It is about 28 times greater than the earth. The distance that light would fall in 4 seconds on the surface of the sun is thus 0.5 x 28 x 9.8 x 4 x 4 = 2195 m or about 2 km. While not nothing, this is once again a very small amount compared to the diameter of the sun. It is so small in fact that I don't believe that it could have been measured photographically in 1900 during the eclipse of the sun. The demonstration in May 1919, by Sir Aurthur Eddington (1882-1944) of light bending by the sun, during a total solar eclipse, was considered to be proof of Einstein's theory of gravity. According to the admittedly, rough and ready figures, that I have shown above, the reported bending seems to be much larger that can be accounted for by gravity alone. This suggests that the extra bending of light is due to bending by other forces such as the powerful magnetic fields that are known to be around the sun. Do I hear a howl of protest that magnets cannot bend light? Well you are probably right and wrong. Light bending by magnetic fields has not been adequately demonstrated on earth, but magnetic fields have been shown to change the plane of polarized light. If they can do this then they must be affecting the path of the photons in some way. If the magnetic fields were very much stronger, or different in nature, then bending might be possible. However, having said this, as I shall mention later I don't believe this is true. There must be something else; besides the accelerating aether flow into the sun (gravity) that is bending the light near the sun.

Let's consider the situation in a really massive star. Let's switch on a hand held torch. A beam of light comes out that is diffusing from the source at the speed of light. Photons are accelerated to the speed of light by the atoms of the filament of the globe by a process which is totally unknown. The acceleration of photons to the speed of light is obviously extremely rapid; nevertheless it is still an acceleration. An accelerating particle causes a pressure gradient to develop in the aether that opposes the acceleration of the particle and this requires energy to overcome. The electricity from our batteries provides this. Once the photon has left the atom it continues through the aether at a constant speed, the speed of light, because there is no more acceleration and no resistance. Now let's consider what happens if we shine our light beam out into space, away from the earth. As the beam leaves the earth it is decelerated by the aether which is accelerating into the earth. Light travels so fast that the aether particles accelerating into to the earth would have very little effect. In 0.1 second the light would have traveled 30,000 km from the earth but deceleration due to aether particles accelerating into the earth would have reduced the speed by less than 1 m/sec. This compared to the speed of 300 million meters/sec, while not being zero is immeasurable. The same applies to the photons that leave the sun and shoot off into space never to return. While the deceleration due to the aether accelerating into the sun is not zero the effect on the speed is so small that it inconsequential.

Pretend that you could stand on the surface of a really massive star, and shine the light from your torch out into space. If the sun you were standing on was thousands of times larger than our sun, then the effect of the aether accelerating into the star would not be that small. If the aether acceleration into the sun was 30,000 m/sec^2 instead of 28 m/sec^2 then there really would be quite a big difference. In one second the speed of light would be reduced by 10% and in the next second it would be further reduced and so on. If the acceleration of the aether into a massive star was high enough the photons would leave the star, decelerate until they were not moving and then accelerate backwards into the star from which they had just left only seconds earlier. This would be a black hole.

Calculating the velocity of the aether

Now that's great you might say. We are pushed onto the earth by aether particles accelerating into the earth, on their way to annihilation by aether destroying particles in the earth. This is gravity and the concept is really pretty simple. Is there any way to calculate the velocity of these particles at any point in time? Can we put a little meat on the bone, and make the theory a bit more solid? Well yes there is a way. It is possible to overcome the drag force in one direction by accelerating at the same rate in the opposite direction. The other concept to consider is that a particle in an accelerating aether flow will accelerate until its speed and acceleration is the same as that of the aether. The corollary of this is that if you know the speed and acceleration of such an object, then you can immediately infer the speed and acceleration of the aether. Bearing in mind of course, that this only holds for long lasting stable and constantly accelerating aether flows, just like gravity.

on the calculations let's summarize our thoughts regarding aether flow.

The principles of aether flow

One

Particles and bodies moving through the aether at a constant speed will continue to move through the aether forever in that direction and at that speed.

Accelerating aether flows will exert a force on, and induce acceleration in particles and bodies in the accelerating aether flow. The acceleration will be in the same direction as the accelerating aether flow.

A to and fro compression wave of the aether surrounds all moving particles whether they are accelerating or not, and this acts at right angles to the direction of motion of the particle. This compression wave will induce acceleration in a second particle at right angles to the direction of motion of the first particle and cause the first particle to bend towards the second particle. The magnitude of bending will be inversely proportional to the square of the distance between the two particles.

Accelerating bodies will exert a force on, and induce acceleration in the aether in the same direction as the acceleration of the body. An aether pressure gradient will develop across the accelerating body and cause a force to act on the body in the opposite direction to the bodies' acceleration. Acceleration of a particle occurring during an impact collision and will cease immediately the colliding particles separate and the colliding objects will thereafter travel at constant speeds.

Two

When aether is destroyed or created aether will move from areas of temporarily high aether density to areas of temporarily low aether density until the density of the aether is equilibrated and the density is a constant. During this movement the aether is accelerating. The pressure of the aether at any point is proportional to the square of the distance to or from the point of aether production or destruction. The acceleration of the aether induced by a pressure gradient is proportional to the difference in aether pressure. In the case of a stable long lasting aether pressure gradient, such as gravity, then the speed of the aether at any point is inversely proportional to the square root of the distance to or from the point of aether destruction or creation. If the acceleration is changing this relationship does not apply. The speed of a particle within an accelerating aether flow will increase until the speed of the particle is the same as the speed of the aether at any point. When this point is reached then the acceleration and speed of the aether and the particle will be the same.

Three

Accelerating or decelerating aether acts on subatomic particles individually. All normal materials will be affected to the same extent by an accelerating or decelerating aether flow because all matter is made of the same subatomic particles. No particle can be moved unless it is in the path of an accelerating aether flow, or unless it is involved in a direct impact collision with another particle.

Four

Aether does not enter the interior of subatomic particles except for the small amount that is being destroyed or created.
The mass of a particle is equal to the number of aether particles that it displaces. The momentum of a particle or body is equal to the product of the number of displaced particles and their velocity where the body is moving at constant speed.
The energy of a body refers to its ability to accelerate aether and is proportional to the number of accelerated aether particles, times the rate of acceleration and the duration of the acceleration.

A particle accelerated as the result of an impact collision will induce acceleration in the aether and an aether pressure gradient will develop across the accelerating particle and will oppose the acceleration. This aether acceleration will only occur only while the two colliding particles are in contact.

All energy whether it is chemical, electrical, magnetic or nuclear has a common pathway ending with the acceleration of aether.

Orbiting bodies

Let's consider a satellite in orbit around the earth. At any point in its orbit there are two components to the velocity which are at play. There is a horizontal component and a vertical component. Once the rocket motors have switched off there is no acceleration in the horizontal direction. The downwards acceleration in the vertical direction is constant, due to the acceleration of aether into the earth (gravity). The speed of an object at any point in a stable accelerating aether flow is proportional to the acceleration of the aether, provided the acceleration is constant, and inversely proportional to the square root of the distance from the aether destroyer.

The horizontal component is parallel to the surface of the earth. The speed of the satellite in the horizontal direction does not change. Since the speed is constant the object is not accelerating. An object traveling through the aether at a constant speed is not affected by the aether. Remember Newton's equation; force = mass x acceleration. Since the acceleration is zero there will be no force acting in the horizontal direction.

The vertical component is acting at right angle to the surface of the earth. The aether is traveling into the earth and hits the surface at right angles. Aether is being destroyed in the earth and that is why the aether enters the earth. The aether is not only going into the earth but it is accelerating into the earth. From the first law of aether motion accelerating aether exerts a drag force on all objects in the accelerating aether flow. That is why objects fall to the earth. If the object accelerates upwards with the same acceleration as the aether, but in the opposite direction, then the object will stop falling.

A low earth orbital satellite travels at about 7.8 Km/sec. This is a circular path. There are two components to its velocity. The one component is upwards, directly away from the center of the earth. The other is in a direction at right angle to the vertical, in the same direction as it travels around the earth. Now let's try to separate these two components. Let's assume that the satellite travels one quarter way around the earth in a circular orbit. Just to refresh your memory the circumference of a circle is 2 x π x r. The distance it has traveled is thus 0.25 x 2 x π x r = 1.57 x r, where r is the radius of the earth. The horizontal component is the radius of the earth. Figure 6 shows a diagram of this concept.

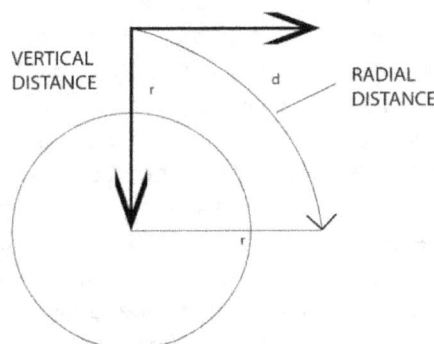

r

d

RADIAL
DISTANCE

r

Figure 6   The concept of horizontal and vertical components of an orbiting satellite

In the diagram the circumferential distance (d) that the satellite has traveled is 1.57 times the vertical distance (r). If we use this fraction applied to the speed we will see that the vertical speed is about 0.6 times the circumferential speed. If the aether is travelling faster than the particles it is accelerating then the aether particles are entering the earth at a minimum speed of 0.6 x 7.8 km/sec or about 5 km/sec. This is about 18,000 km/hour. Now we need to go back and consider how an object moves through the aether.

If gravity is indeed a drag force due to accelerating aether, is it possible to accurately calculate the speed of the aether as it passes into the earth? This might sound an easy task, just work out the vertical component of an orbiting body and that's your answer. Well that's what I thought myself but later I realized that it is not that simple. The problem is that in a frictionless environment the intuitive solution is incorrect, because even a tiny acceleration can accelerate an object to incredible speeds if left long enough. In the world we are familiar with, and on which our intuition is based, the speed of an object is always limited by friction. This just does not apply to the aether and subatomic particles.

Escape velocity versus escape acceleration

There was a lot of confusion in my mind about escape velocity. Was this referring to a speed straight upwards, or a speed parallel to the ground? I know that if I have a problem with a concept, then so do hundreds of millions of other people.
Let's consider a rocket blasting off into space. When the rocket is accelerating at 9.8 m/sec^2 the gantry can be removed and the rocket will remain stationary in a vertical position. The engines would be firing furiously, consuming fuel at a tremendous rate, but the rocket would just sit there suspended between heaven and earth. If we increase the power of the motors then the rocket would slowly leave the launch pad and accelerate upwards into space. If we leave the engines running long enough it will never come back. Now am I saying that the rocket will go into space without achieving escape velocity? That's correct. What we need to achieve is escape acceleration, not escape velocity. As long as the rocket is accelerating faster than 9.8 m/sec^2 it will go straight up into space. It does not matter how slow it goes it will eventually get into space. If the rocket was accelerating at 9.9 Km/sec^2 you could stand on the

ground and wave good bye to the astronauts standing at the windows inside the rocket. They could be sipping cocktails and chatting to you on their cell phones, as the rocket slowly leaves the launch pad and gradually rises into space. This would be an elegant Hindenburg type departure. So why don't rockets take off like this? What you actually see is the astronauts strapped in their seats, being pushed down with high g forces, while the craft shakes and vibrates furiously. This is hardly the gentlemanly departure I portrayed above. I think it is primarily because of the difficulty in controlling the power output of the rocket motors. The design of rocket motors has not really changed since their invention during World War II. In the Science Museum in London there is a superb collection of rocket motors. Each one is bigger than the one before but they all look the same just larger. At the heart of the motor is a positive feedback mechanism. As the exhaust pressure rises, this pressure is used to pump more fuel into the rocket, which in turn causes the pressure to rise still higher. It is similar to the turbo charger on a car. This design enables massive volumes of fuel to be pumped in a very short time. This is a great design for bombing London, but it is not so good for future wealthy elderly space tourists with dicey hearts. Also remember that the shuttle uses solid fuel boosters, and once these are ignited they cannot be controlled at all. Another factor is probably confused thinking by the engineers, who are muddling up speed and acceleration, though of course they will furiously deny this allegation.

Now let's consider the speed of the aether in these situations. Because the aether pressure gradient produced by the earth is constant the acceleration due to the pressure gradient is a constant at any point and the force acting on the aether particles is also constant at that point. The speed of the aether at any point along the aether pressure gradient will be proportional to the acceleration at that point. This brings us to an important conclusion. Where there is a stable aether pressure gradient such as gravity, then the speed of the aether at any point along that gradient will be proportional to the square of the distance to the aether destroying object.

To work out what is going on when a satellite is in orbit we need to separate the horizontal from the vertical components of the velocity. As discussed before. As mentioned the relationship of the radial to the vertical component is 1.573 to one. This relationship between the vertical and the horizontal comment holds for an orbit of any radius. The orbiting speed of a low earth orbiting satellite is 7.8 km/sec. If we apply our factor to this the upwards speed is 7.8/1.573 or about 5.1 km/sec. Since the satellite is not falling, and the acceleration is constant, we can say that the downward acceleration of the aether particles are at least equal to the upwards acceleration of each of the subatomic particles making up the satellite.

Why is the earth in orbit around the sun?

Why does the earth orbit the sun? Because the sun is destroying aether within its interior the pressure of the aether around the sun would fall in relationship to the aether in the surrounding space. Aether then moves from temporary high aether density to temporary low aether density so that the density of aether is maintained at a constant. The movement takes place because aether is basically incompressible. Because aether is incompressible, as the aether moves into the sun it accelerates. The further we move away from the sun the smaller the acceleration. As we get closer to the sun the aether acceleration increases. The accelerating aether exerts a drag force on the planets. We know this drag force as gravity. We also know that the size of the sun has been a constant for a very long period. In the situation where the acceleration has been a constant for a long time, then the speed of an object at any point can be calculated. In the case of the planets we know the starting speed of the aether far out in space, namely zero. We know the distance of the planet from the sun and we know that the acceleration of the aether is not changing. Under these circumstances we can say that the vertical component of the velocity of the planet around the sun would be proportional to the square root of the distance of the planet from the sun. The speed of the aether moving towards the sun decreases by the square of the distance from the sun. If we double the distance to the sun then the aether is moving 4 times slower at that point. If we double the distance again then the aether is moving 8 times slower and so on.

How to lose weight really fast

A very fast and sure way to lose weight is to take a vacation to the equator. What, you don't believe me? No it's not the tropical diarrhea that will cause the weight loss it's the centripetal force of the earth. As the earth turns on is axis once every 24 hours you would be moving at about 1647 Km/hr on the equator, due to this rotation. However it is 1278 Km/hr in New York City. If you had to travel to the poles the speed there is zero. Because of this speed you would be flung off into space if it were not for gravity pushing you onto the surface.

The equation which gives this centripetal acceleration is

$$a = v^2/r$$

where a is the acceleration in m/sec^2, v is the velocity in m/sec you are travelling in the direction horizontal to the earth, and r is the radius of the earth in m. If you do the calculation you will see that this acceleration is very small in relation to the acceleration due to gravity, but it is not zero. You will weigh less at the equated than in New York. To convert this centripetal acceleration into force we multiple the acceleration by the mass of the object. $F = mv^2/r$ where m is the mass of the object of interest.

What is the relationship between the speed of the planets and the speed of the aether at any point?

This requires considerable thought. The force of gravity on the planet basically depends on the aether pressure gradient across each quark making up the planet, and the combined surface areas of all the quarks. This depends on the difference in the speed of the aether across the width of each quark and on the surface area of each quark. Taking both factors to the extreme we can see that if there was no difference in aether speed there would be no force, and if the surface area of the quark was zero there would also be no force. However we cannot easily get the speed of the aether from the speed of the quark. Let's take an airplane wing as an analogy. During flight the air travel further over the top of the wing than underneath the wing because of the curved shape of the wing. Because the air is 'thinned out' over the top of the wing the pressure is lower at the top of the wing than underneath the wing. Because of the pressure difference there is an upwards force on the wing which gives the wing lift. This keeps the airplane in the air. Now what figures can we deduce. Firstly we know that the upwards force is causing an upwards acceleration on the wing, and therefore on the whole plane. What is this acceleration? Well, it must be close to 9.8 m/sec^2, the acceleration due to gravity. If it was bigger than this the plane would go up and anything less than this and the plane would start to fall. If we did not know what the airspeed was and wanted to work this out I am sure clever engineers could do this because, they know the surface area of the wing, the upwards force on the wing and the weight of the airplane. Now when we come to the planets in orbit around the sun and we try to work out the speed of the aether it is analogous to working out the airspeed necessary to keep a plane in the air. However when it comes to working out the speed of the aether we have a problem, we don't know the surface area of the quarks. All we know for sure is the acceleration of the aether. We know this from the centripetal acceleration of the planets.

Now let's return to the wing analogy to discuss another problem with determining the speed of the aether. Let's say we magically remove friction, and at the same time we make the weight of the plane zero, but keep the engine running normally what would happened? Well the plane would rise because the weight of the plane has been removed. The plane would accelerate upwards at 9.8 m/sec^2. The plane will rise and continue to rise. The horizontal speed would not change because the engine is running at a constant speed. The horizontal speed is 1000 Km/hr and the vertical upwards speed would initially be very small in relation to that, maybe on a few Km/hr, but it is accelerating. Because there is no friction the upwards speed will continue to increase relentlessly as long as the plane is moving forwards. At some point the upwards velocity would be higher than the horizontal air velocity and it would still be accelerating.

When it comes to planets our intuition tells us that the vertical speed of the aether must be going faster than the vertical speed of the planets but, as this wing analogy shows this that might not be true either. This is because the aether is operating in a frictionless environment. Let's consider an object where the aether is travelling at 1 km/sec on the one side of the object and 2 km/sec on the other side. There will be a force on the object of say x Newton. Now consider the same object where the aether speed on the one side is 1000 km/sec and on the other side the speed is 10001 km/sec. The acceleration across the object will be the same as the first case and the force will also be x Newton. This is very counter intuitive. We would immediately believe that the force would be much greater in the latter case, but in a frictionless environment only the difference in the speed matters. This concept lies at the heart of this gravity theory.

Trying to relate the acceleration of the planets to the speed of the aether is a formidable task, and far beyond my abilities, but logic tells me that there must be a relationship. I propose that the aether speed at a particular distance from the sun is proportional to the vertical component of the planet's velocity at that5 distance.

$$V_e = k \times V_p$$

Where V_e is the speed of the aether in Km/sec, V_p is the orbital speed of the planet in Km/sec and k is a constant the value of which we do not know. How could we find out what this relationship is? I think the answer will come from the field of nuclear physics when the strong nuclear force is related to the aether acceleration into the atom. Once the relationships between the surface area, mass

Proof of the inverse square law

As the earth orbits the sun it tends to fly off into space due to its centripetal acceleration. This is like a bucket that you swing around tethered by a string. The bucket is kept in place by the force of the string on the bucket. This force pointing to the center is called the centrifugal force and the force that flings you outwards is the centripetal force. As you swing the bucket the centrifugal force, which acts towards the center, exactly equals the centripetal force and the distance of the bucket from you does not change. If you cut the string, the centrifugal force fall to zero and the bucket will fly outwards. This is a force, and as you know all forces are caused by acceleration. It is the centripetal acceleration that causes the centripetal force. In the case of the earth in orbit around the sun the sun, there is no string holding the earth to the sun rather the force that balance the centripetal force is gravity. Thus the centripetal force is exactly counterbalance by the force exerted by the accelerating aether flow. For any planet in orbit we can say that the centripetal acceleration acting on the planet is exactly balanced by the acceleration caused by the aether drag.

Centripetal force = gravitational force
Centripetal acceleration x mass = aether acceleration x mass

Since the mass is the same we can say that
Centripetal acceleration = aether acceleration.

To prove the inverse square law we need to show that at any point in space the centripetal acceleration is inversely proportional to the square of the distance to the aether destroyer.

The aether acceleration at any point in space would be given by the equation $a_e = v^2/r$ where a_e is the acceleration of individual subatomic particles such as quarks caused by the acceleration of the aether towards an aether destroyer in m/sec^2, v is the angular velocity of an object in orbit around the aether destroyer, and r is the distance to the center of the aether destroyer. Notice that the relationship is between the aether acceleration and not the aether speed.

I shall now demonstrate this relationship using the planets. Now before you go off and think I am some sort of genius for working out these equations this is not true. The relationship between the orbital speed, the centripetal acceleration of the planets, and the distance to the sun has been known, in one form or another for hundreds of years.

Where my work differs is that I use the concept of the aether to explain the relationship. The first postulate is that for any planet the aether acceleration which causes gravity, is exactly balanced by the centripetal acceleration acting on the planet. The second postulate is that the aether acceleration at any point is inversely related to the square of the distance to the sun.

The aether acceleration towards any object would be given by the equation is $a_e = Z/r^2$

Where Z is a constant, a_e is the aether acceleration at any point in m/sec^2, and r is the distance to the object in meters. The value of Z depends on the amount, and rate of aether destruction in the object.

Now if the centripetal acceleration is exactly balanced by the aether acceleration then,

$v^2/r = Z/r^2$

We want to know the value of Z. Rearranging the equation we get

$$Z = \frac{v^2 \, x \, r^2}{r}$$

$$Z = v^2 r$$

r is the radius of the orbit of the planet. Z is a constant for the sun and is determined by the mass of aether destroyed in the sun per unit time. The aether destruction is determined by the excess number of aether destroying particles over the aether producing particles. v is the orbital velocity of the planet.

The units of Z are m^2 x m / sec^2 = m^3/sec^2. We will just call them units/sec.
Now let's build a table based on astronomical observations of the planets in order to determine the value of Z. We will exclude Pluto because the orbit is so eccentric that there is no reliable average speed on the internet.

Having read this paper so far I am sure that you will be the type of person who will want to confirm these calculations and apply them to other moon systems. Let me give you a few tips before you start. Be careful with the units. I have reduced everything to meters and seconds. It gives you really big numbers but it avoids making mistakes. Don't forget that when you are dealing with the orbits of planets you are dealing with millions of Km, whereas you are dealing with thousand of Km when you are dealing with moons.

The constant Z will have the units U/sec. Units being the net amount of aether destroyed by the sun in one second.

Planet	Orbital Velocity (v) m/sec	v^2	Distance from the sun (m)	Z (v^2r)
Mercury	47.9×10^3	2294×10^6	58×10^9	1.33×10^{20}
Venus	35.0×10^3	1225×10^6	108×10^9	1.32×10^{20}
Earth	29.8×10^3	888.0×10^6	150×10^9	1.33×10^{20}
Mars	24.1×10^3	580.8×10^6	228×10^9	1.32×10^{20}
Jupiter	13.1×10^3	171.61×10^6	778×10^9	1.33×10^{20}
Saturn	9.6×10^3	92.16×10^6	1430×10^9	1.32×10^{20}
Uranus	6.8×10^3	46.24×10^6	2880×10^9	1.33×10^{20}
Neptune	5.4×10^3	29.16×10^6	4504×10^9	1.31×10^{20}

Table 2   The relationship between the orbital velocity of the planets and the distance to the center of the sun.

The figures are very close. The difference is probable more to do with rounding errors than incorrect measurement.

Taking the average value of a_e you can see that the sun has a Z value of 1.32×10^{20} U/sec.

If we make the assumption that the aether is travelling faster than the vertical velocity of the planet then using the factor of 1.57, discussed above, we see that the minimum speed of the aether as it passes the earth is 19.0 km/sec whereas it is a minimum of 30.5 km/sec as it passes Mercury. This proves that the aether is indeed accelerating.

Let's take the radius of the sun as roughly 695,000 Km and plug the values into the equation in order to determine the minimum aether speed as it goes into the sun at the surface.

$$V = \sqrt{\dfrac{Z}{r}}$$

$$V = \sqrt{\dfrac{1.32 \times 10^{20}}{6.95 \times 10^{8}}}$$

$$V = \sqrt{0.199^{12}} = \sqrt{19.9 \times 10^{10}}$$

$$V = 4.47 \times 10^{5} \text{ m/sec}$$

This would be the speed of a planet at the surface of the sun. To get the minimum vertical speed of the aether we have to divide by 1.57. This gives us a minimum speed of 2.85×10^{5} m/sec or 285 Km/sec.

As we go further out we see that the speed of the aether falls of dramatically. One light year is a distance of about 9.46 trillion Km. Let's plug this value into our equation.

$$V = \sqrt{\dfrac{Z}{r}}$$

$$V = \sqrt{\dfrac{1.32 \times 10^{20}}{9.46 \times 10^{15}}}$$

$$V = \sqrt{0.14 \times 10^{5}} = \sqrt{14000}$$
$$V = 118.3 \text{ m/sec}$$

Let's convert this to Km/hour to make it more meaningful.

$V = (11.8 \times 3600) / 1000 = 42.5$ Km/hour

Objects orbiting the earth at one light year would be orbiting at this speed, which is about the speed you commute to work. The aether speed at this distance would likewise be very slow at this distance. You can obviously see, that for the aether to cover a trillion Km to the sun it is going to take a little while at this speed. The outer edge of the Ort cloud, which is the furthest region of our solar system, is thought to be about one light year from our sun.

Now let's apply the same reasoning to the earth. I have only three points to demonstrate the equation, namely the moon, a geostationary satellite, and a near earth orbiting satellite. Notice that once again the values fit the equation.

	Orbiting velocity (m/sec)	V^2	Distance to center of earth (m)	Z
Low earth orbiting satellite	7,800	60.8×10^6	6.47×10^6	3.93×10^{14}
Geostationary satellite	3,070	9.42×10^6	42.26×10^6	3.98×10^{14}
Moon	1,020	1.04×10^6	384.0×10^6	3.99×10^{14}

Table 3   The relationship between the orbital velocity of earth satellites and the distance from the center of the earth

In the case of the earth the Z value averages 3.97×10^{14} U/sec

I have done the same thing with 4 of the moons of Jupiter. Once again the figures fit fairly closely with the equation.

	Orbiting velocity (m/sec)	V^2	Distance to center of Jupiter	Z
Io	17,300	299.3×10^6	493.1×10^6	147×10^{15}
Europa	13,700	187.7×10^6	742.4×10^6	138×10^{15}
Ganamede	10,900	118.8×10^6	1141.5×10^6	136×10^{15}
Callisto	8,200	67.2×10^6	1951.5×10^6	131×10^{15}

Table 4   The relationship between the calculated aether velocity passing the moons of Jupiter and the distance to the center of Jupiter

In the case of Jupiter the Z value averages 1.38×10^{17} U/sec
How do these Z values relate to the mass of the object, using the conventional definition of mass?

Object	Mass Kg	Z	Mass / Z
Sun	1.99×10^{30}	1.32×10^{20}	1.51×10^{10}
Jupiter	1.90×10^{27}	1.38×10^{17}	1.38×10^{10}
Earth	5.98×10^{24}	3.97×10^{14}	1.50×10^{10}

Table 5   The relationship between the Z values to the mass of the body

As expected the results are pretty close to a constant. The destruction of the aether in the sun or a planet is proportional to the excess number of aether destroying particles in the body in question. Because all matter is made of the same components there will be a close relationship between the number of aether destroying particle to the total number of particles. The mass of the object is a proxy for the total number of aether destroying particles in the body.

The above tables seemed pretty neat and tidy so I got a bit ambitious and tried to apply it to the galaxy. The main problem is that I only have one point, but that may be enough to determine the Z value of the galactic center. The earth is thought be 27,500 light years for the center of our galaxy. After many years of theoretical speculation there is now very strong photographic evidence that there is a super massive black hole at the center of our galaxy. It is thought that these are probably found at the center of every galaxy. What we want to know is the Z value of the galactic center.

Let's use the equation $Z = rV^2$ to do the calculation.

The orbital speed of the sun around the galaxy is thought to be 30 Km/sec. Later I will explain why I doubt this is correct. But for now let's assume that it is correct. Let's plug the figures into our equation. There are 9.46×10^{12} Km in a light year. We bring everything to meters as before/

$$Z = rV^2$$

$$Z = (27{,}500 \times 9.46 \times 10^{12} \times 1000) \times (30{,}000 \times 30{,}000)$$

$$Z = (260{,}150 \times 10^{15}) \times 900 \times 10^6$$
$$= 2.60 \times 10^{20} \times 9 \times 10^8$$
$$= 2.25 \times 10^{29}$$

Now the Z value of our sun is 1.32×10^{20}. The ratio of the galactic center to our sun is 1.70×10^9 to one. This means that the galactic center is destroying roughly 2 billion times more aether than our sun in the same time period. If, and this is a very big if, the nature of matter within a black hole is the same as normal matter then we can say that the galactic center should be about 2 billion solar masses. Now this presents a very significant problem. Using Newton's law of gravity the mass of the center of our galaxy has been calculated at 3.7 million solar masses. There is a big discrepancy here. I get a figure 475 times greater.

If we know the Z value and we know the radius of the galactic center we can work out the aether speed at the surface of what is thought to be a black hole. Our sun has an aether speed of 283 Km/sec at the surface which is a very small fraction of the speed of light. The most accurate measurement of the diameter of the center of the galaxy has been made with the Very Long Base Array which is a series or very large radio telescopes. They have measured the diameter as 93 million miles which is almost the same as the distance of the earth to the sun. This is equal to a radius of 75 million Km.

If we plug this value in the equation then we get

$$V = \sqrt{\dfrac{Z}{r}}$$

$$V = \sqrt{\frac{2.25 \times 10^{29}}{75 \times 10^9}}$$

$$V = \sqrt{0.13 \times 10^{20}}$$

$$V = \sqrt{13 \times 10^{18}}$$

$$V = 3.60 \times 10^9 \text{ m/sec}$$

$$V = 3.6 \times 10^6 \text{ Km/sec}$$

This is the orbital speed of a planet at the surface of the center of the galaxy. To get the vertical speed we divide by 1.57. This gives us a speed of 2.29×10^6 Km/sec

The speed of light is roughly 3×10^5 Km/sec. These figures indicate that the minimum speed of the aether entering the center of the galaxy at slightly over 7 times the speed of light. This would strongly suggest that the center of the galaxy is indeed a black hole. However if the published figure of 3.7 million solar masses is used instead of my figure of 1.7×10^9 solar masses then we get a very different picture. If the mass of the center of the galaxy is 459 times smaller than my figure then the Z value would be $2.25 \times 10^{29} / 459 = 0.0049 \times 10^{29} = 4.9 \times 10^{26}$ U/Sec. We get the following result

$$V = \sqrt{\frac{4.9 \times 10^{26}}{75 \times 10^9}}$$

$$V = \sqrt{0.065 \times 10^{17}}$$

$$V = \sqrt{65 \times 10^{14}}$$

$$V = 8.06 \times 10^7$$

$$V = 806,000 \text{ Km/sec}$$

The vertical speed of the aether would be 513,000 Km/sec

This is still 1.7 times the speed of light and once again it suggests that the center of the galaxy is a black hole. If these figures are correct, and the center is a black hole you may well ask how the aether can go faster than the speed of light because I went to great length earlier to show that the speed of light is a constant. Well it still is. Remember that the speed of light is a constant because the density of the aether is a constant. The only way to keep the density of the aether a constant is for it to go faster. The speed of light is a constant not the speed of the aether. If you are inside the aether flow you will not know the speed of the aether because you are surrounded by aether moving at the same speed. Particles such as light will still behave in the normal manner, even thought the aether flow is going considerably faster than the speed of light.

Now notice something else. We can determine the speed of the orbit of an object at any position in space with respect to the earth, and the sun, and the center of the galaxy and we do not need to know the mass of any of the objects concerned, or their size, or shape or composition. Having been brought up on Newton's equation for gravity I am sure you will find this a little hard to swallow. How can this be so? Well it is because every object in the universe is made of the same subatomic particles and the aether acts on the subatomic particles as if they are individual things. The earth in orbit around the sun is basically in the same position as Galileo's balls falling from the top of the tower of Pisa. Just as his balls all accelerate at the same speed towards the earth, the earth is always accelerating towards the sun. Galileo's balls hit the ground, but the planets in orbit, just fall forever. It does not matter what size or shape or composition the planets are they are all regarded by the accelerating aether as being individual particles.

The effect of multiple bodies, and bodies of equally large size

You probably have noticed that I only talk about the effect of the accelerating aether flow caused by the large body on the smaller one, and totally ignore the effect of the accelerating aether flow into the small object. The reason for this is that in the case of all the examples I have given, with the possible exception of the moon and the earth, the large object is so large in relation to the smaller one that the effect of the smaller one on the larger one is basically negligible. Though Jupiter is our largest planet it is very tiny in relation to the sun. Don't think in terms of diameter, but in terms of volume. This is of course a cubic relationship. However to work out the accurate speed that two planets would approach each other, if they were not in orbit, you would have to calculate the acceleration of the aether into the one object using the Z value of the other object, and vice versa. In the case of three or more bodies the same principles would apply but you would also have to take into account the angles between the planets. This will rapidly become extremely complex problem and well beyond my capabilities to solve. In Newton's law of gravitation you notice the product of the masses is used and not the sum of the masses. Why is this? The mass of an object is an approximation for the number of aether particle being destroyed per unit time, and because this is the cause of gravity, it is no surprise that this is directly proportional to the force of gravity. If you have two bodies and you double the mass of one of them you will double the force of gravity between them. If you double the mass of the other body you will double the force again and end up with four times the force. If you treble the mass of one of the bodies you will treble the force between the bodies. If you treble the masses of both bodies you will treble the force and then treble it again ending up with nine times the force. The masses are thus multiplied and not added.

Where does the energy come from to keep accelerating the earth forever? Well it comes from the destruction of the aether in the sun. It is the continuous force of creation and destruction that is responsible. Let's take practical example. If I give you a distance from the sun of say 150 million Km, which is actually that of the earth, and the Z value of the sun, then you can calculate the speed that the object has to be moving to be in orbit. For a body, any body, to be in orbit at that point, it will have to be moving at a speed of 29.8 Km/sec at right angles to the sun. If it is too fast it will shoot off into space, if it is too slow it will crash into the sun. Now I said any body. It can be a planet as big as Jupiter, or a body the size of a pea or a gas or anything else, it makes no difference. The only thing that matters is the speed of the aether, which is proxy for the acceleration of the aether, as it passes through the planet, not the mass of the object, because all normal matter behaves the same in an accelerating aether flow.

Planet formation and the density of planets

What I said in the above paragraph really sounds strange. You cannot easily imagine a planet like Jupiter having the same orbit as the earth. As you know the planets close to the sun are dense and stony, while the planets father out are gaseous, with a much lower density. If I am correct then why would you not find gaseous planets close to the sun and stony planets further out? I speculate that when the planets formed from the disk of matter and gas around the sun the acceleration required to get to the correct orbiting speed was important. The closer the planet is to the sun the faster it moves. It moves faster because the aether acceleration is greater. For it to move fast it must have been subjected to a significantly higher acceleration than if it was further away. The only planets that would survive a rapid acceleration would be the dense planets because the atomic forces holding the planet together are greater than the forces holding a gaseous planet together. At a distance of one light year from the earth the picture is very different. I suggest that this is why Schumacher Levy 9 broke up when it got near Jupiter. Like all comets this comet had formed far out in space where the angular velocity is very small. The acceleration required to get to the orbiting velocity at that distance was very small and even an object made of gasses would hold together. When the comet got near to Jupiter the situation was very different. The comet was very unlucky. As it tried to sneak across the orbit of Jupiter it found itself in the wrong place at the wrong time and basically got hit by the planet. As Jupiter approach the comet from the side, the comet was subjected to two simultaneous accelerating aether flows, at right angles to each other, one caused by the sun and the other caused by Jupiter. Since it had formed in a region of very low aether acceleration, it was simply not able to take the forces causes by the two opposing accelerating aether flows, and it broke up into pieces before it crashed into Jupiter. I postulate that during the formation of the solar system a similar thing happened to all the forming planets, and that the only planets that could withstand the more powerful accelerating aether flow near the sun were the planets made of stone and iron. It is a bit like planet evolution, the weaker planets got destroyed and disappeared in an environment of high aether acceleration and only the tough stony and metallic ones survived near the sun.

The new universal constant

Though Einstein proposed that the speed of light is the universal constant I propose that this is not entirely accurate. The speed of light is dependent on the density of the aether, and this in turn is related to the rate of destruction of aether by matter. I propose a new universal constant which is the rate of aether destruction per unit time per subatomic particle. This theory is based on the premise that aether is destroyed in aether destroying particles, most probably quarks, at a constant rate. Some aether might be made in other subatomic particles, but overall the rate of aether destruction for every atom is a constant and is related to the number of subatomic particles in that atom. This rate of destruction appears to be a constant. Overall one fixed amount of matter will destroy the same number of aether particles in a given time everywhere in the universe. I will call this constant E.

This is the universal law of aether destruction

The number of aether particles destroyed per unit time in an object of mass m equals the universal rate of aether destruction times the mass of the object.

If we assume that all aether particles have the same mass then we can rephrase the law as follows.

The mass of aether destroyed per unit time by an object is proportional to the mass of the object.

$$n_e = E \times m$$

Where n_e is the mass of aether particles destroyed per unit time, E is the universal aether destruction constant and m is the mass of the object.

Putting units to this we could say that one aether unit is the mass of aether expressed in Kg destroyed in one second by one Kg of matter. This is a universal constant. This can be used to define an aether unit. One aether unit is the amount of aether expressed in Kg that will be destroyed by one Kg of matter in one second.

The units of an aether unit would be $\left(\dfrac{Kg}{Kg} \right) \times \dfrac{1}{sec} = sec^{-1}$

The gravitational force between two objects would be proportional to the mass of first object times the universal aether destruction constant, multiplied by the mass of the second object times the universal aether destruction constant, divided by the square of the distance between the objects.

What goes up must come down?

I remember when the first satellite went into orbit. It was called Sputnik I and was launched on 4 October 1957. It was a shiny sphere about one foot in diameter with radio antennae sticking out from the sides. Nobody had heard of a satellite before. Nobody could explain what it was or how it stayed up. Then one of the teachers suggested that we should just think of it like the moon, but a bit smaller. Oh well, then it was easy. The old adage - what goes up must come down, appeared not to be true. If you could make the object go fast enough it goes into orbit and never comes down. We were given the usual pictures of cannons firing cannon balls, and the cannon ball falling to ground after a short distance. But at higher speeds the cannon balls fell off the edge of the earth and fell forever. It seemed a little strange to me at the time because the direction of a satellite was always changing and surely you would need to keep the rocket engines running to change the direction. I remember a few years later having a discussion about orbiting satellites and mentioning that NASA was somehow "cheating" by getting the satellite to go into orbit. However, the experts seemed to be happy with the "cannonball explanation" and I guessed they knew what they were talking about. Now I realize that my initial skepticism, was justified and the local 'experts' didn't know what they were talking about. It does indeed take a continuous source of energy and lots of it the keep a satellite in orbit. But, the energy is not coming from the satellite or the rocket it is coming from aether destruction in the earth. Let me elaborate on this. Newton first law of motion is that an object will continue to move in a straight line unless an external force acts on it. Since an orbiting satellite is continuously changing direction then there is continuous force acting on it. Sure this is gravity. Now let's take this one step further. Work is force x distance and work is the same as energy. The energy expended to move an object is the force times the distance. Now force is mass x acceleration. The energy expended to move something is thus mass x acceleration x distance. Now if we use the calculation of the vertical distance that I used above let's see how much power it needs to keep a satellite in orbit.

Let's use the Sputnik 1 satellite as an example. The weight of the satellite was 86.3 Kg and it took 98 minutes to complete one orbit. Because it was very close to the earth the acceleration due to gravity is roughly the same as g i.e. roughly 9.8 m/sec^2. Let's assume that the satellite goes straight around the equator. How far would the satellite fall in the vertical direction? Well if we can use our equation used above and divide the circumference by 1.57. The circumference of the earth at the equator is roughly 40075 Km. The distance a satellite would 'fall' in one orbit would then be 40075/1.54 = 26022 km. Now let's plug in the values, remembering to convert Km to meters. The work done would be 86.3 x 9.8 x 26022 x 1000. This equals 22,007,846,280 Joules. The units are in Joules. This does not mean that much to us. Let's rather see how this relates to power consumption. Kilowatt hours are something that we understand because we pay for this every month in our utility bills. Power is the work performed per unit time. One watt is equal to one joule per second. Know we need to know how long we have been applying this force on the satellite in order to complete one orbit. In this example the force was applied for 98 minutes or 5880 seconds. The power consumption is thus 22007846280/5880 = 3,742,831 watt. Now let's work on this figure a bit to make it more meaningful. This is roughly 3.7 megawatt. The power output of a super car such as a Ferrari is 0.55 MW (747 hp). The power output of the fastest Second World War propeller driven fighter, the P51 Mustang was 1.3 MW and a standard diesel electric locomotive on the railways is about 3 MW. Let's put a cost on this. The average cost of electricity to consumers in New York is roughly 20c per kilowatt hour. If you were paying to keep Sputnik 1 in orbit it would cost you $740 per hour. This is about the same amount as a second rate divorce lawyer charges.

You can see that the energy that the Russians used to get the satellite into orbit is tiny compared to what is needed to keep it in orbit for any length of time. Remember that we are talking about the very first and smallest Russian satellite ever launched. This is a lot of power. Now think about how much power is needed to keep the moon in orbit around the earth. Where is all the energy coming from? It is coming from the creation of aether and its ultimate destruction within the earth.

Michelson and Morley's experiment revisited

Now that you know the laws of aether motion we can return to this famous experiment and explain why they got a negative result and point out the error of Einstein that ultimately lead to his ridiculous theory of gravity, namely that it is this due to the bending of space-time. If like me you have had difficulty understanding his theory it is, in my opinion, because it is plain wrong. Not only is the whole notion of space-time ridiculous, but gravity is not even an attraction.

The proposition was that if you measure light in the direction of the earth's orbit the speed would be faster than if you measured it at right angles.

They had light travelling across the gap between two mirrors. They postulated that light traveling across the aether flow would be deflected by the drag of the aether. This deflection could be measured by the generation of an interference pattern. When they got a negative answer they thought that the aether might be stuck to the earth, in some way, They repeated the experiment on top of a high mountain, but got the same result, Others repeated the experiment and verified their findings, namely that the speed of light is the same in all directions. Now where is the error? It is quite simple. They did not realize that only accelerating aether can exert a force. They assumed that light would behave the same as the swimmers crossing the Rhine that I alluded to earlier. They assumed light would be deflected by the aether flowing past the earth as it orbits the sun. Since the angular speed of the earth is a constant the acceleration is zero. From the equation force = mass x acceleration there could be no force because the acceleration is zero. No force equals no deflection of light. Einstein must have had a bad day to fall for this one. Now what about the speed of light in the direction of the sun? Surely this would be different because here the aether is accelerating. The explanation lies in why the aether is accelerating. It is accelerating in order to keep the density of the aether constant. It is precisely because the aether accelerates towards the sun that the density of the aether remains constant. The speed of light is limited by the density of the aether through which it moves. This density is a constant; therefore the speed of light is a constant. That is why even if you measure the speed of light while pointing your measuring device directly at the sun, or away from it for that matter; you will get the same result. The only way you could measure a difference in the speed of light is if you could accelerate very rapidly while taking the measurement. A force would then be exerted on the aether, and for the time that you are accelerating the aether would not be in equilibrium and you might be able to get a different reading.

Does a magnetic field bend light?

This is quite an important question because if gravity bends light and a magnetic field also bends light then there could be a direct link between gravity and magnetism, and this could also explain, what I consider to be, the excessive bending of light by the sun as mentioned earlier. Many experiments have been done to find this out. At very high magnetic fields strengths some have claimed to be able to demonstrate this, but having read the literature to the best of my ability I think the answer is that magnetic fields do not bend light. If there is an effect it is so tiny that it could be due to experimental error. If you think about the very powerful magnetic fields on the surface of the sun and how they move about you would expect to see some effect on the light coming from the sun, such as blurring of the surface, but you don't see this. The next time you are casually flipping through a really thick astronomy book, doing some light reading, have a good look at the detailed pictures of solar flares and sunspots and you can see that everything is in sharp focus. The super strong magnets that are used in medical imaging are also unable to bend light and the pictures are crystal clear with no sign of blurring. That light is not affected by magnetic fields is quite interesting, because photons seem to be intimately related to electrons, being emitted when electrons change orbitals. It has also been known since Maxwell's time that magnetic fields change the plane of polarization of light.

Does an electrostatic field bend light?

The question is similar to the one above. Once again some people claim to have demonstrated this, using massive electrical charges, but overall the opinion of most people is that this is experimental error, and that electric charges cannot bend light.

Gravitational lenses

Using massive terrestrial telescopes, as well as the Hubble telescope, astronomers have observed that some distant galaxies appear to bend light around them. It is thought that this is bending is bending around a black hole in the center of galaxies. These pictures are certainly very impressive and it certainly looks like very convincing evidence that light is bent by gravity. This of course by now is no surprise to you. Could anything else be causing this effect? Well I suppose it is possible that something else could be causing the light to bend such as a gas. It cannot be an artifact in the telescope as several different instruments have taken the same picture. A gas is very unlikely to be causing the bending, as gasses have very little effect on light and this bending is very large.

Why is the moon moving away from the earth?

Ever since a laser reflector was placed on the moon very accurate measurements have been made of the distance from the earth to the moon. It has been demonstrated that the moon is moving away from the earth at the rate of approximately 3.9 cm per year. This might not sound very much but it is quite a large amount taken over millions of years. When you ask people why this is happening the usual answer you get is that the moon must be slowing down because of space dust friction, and just drifting away. Others have suggested that the friction of the tides is responsible and that this loss of energy is causing the moon to slow down and drift away. These explanations are of course completely wrong. If the moon were slowing down say from space dust friction or the tides then it would crash into the earth not move further away. For the moon to move further away according to classical gravity theory, as well as mine, you would have to add energy, and a lot of it, to achieve this movement. Where could this energy be coming from? There is no evidence at all that energy is being added to the moon. If you are not adding energy then the earth would have to be getting lighter. If this was happening then the earth would be moving away from the sun. As it happens there is evidence that this is indeed also happening. The distance for the earth to the sun has been measured with great accuracy and is 149,597,870.696 Km. Now you might be a little skeptical, as I am, about this level of accuracy. Nevertheless Krazinsky and Brumberg have calculated that the earth and sun are in fact moving apart at the rate of about 15 cm per year. So if the moon is moving away from the earth and the earth is moving away from the sun what is going on? For this to be happening, the total energy in the system would be increasing.

Now let's use our aethereal gravity theory to address this problem. The first thing is to realize is that if the moon is moving away from the earth then the centripetal acceleration given by v^2/r is decreasing. Since the centripetal acceleration is exactly balanced by the aether pressure gradient across each particle making up the moon the moon we can say that the aether pressure gradient with respect to the earth must be decreasing. The force of gravity is due entirely to this aether pressure gradient and nothing else. The aether pressure gradient in turn depends on the density of the aether in the space around the earth and the amount and rate of destruction of aether in the earth.

Since the aether density equilibrates to a constant throughout our solar system the effect could be due to decreased aether destruction in the earth. Do you mean the mass of the earth is decreasing? No, the mass of the earth is the number of aether particle displaced by the earth. Gravity is dependent on the number of aether particles being destroyed by the earth. The number of aether particle being destroyed is closely related to the total number of particles in the object, and hence it's mass, but they are not the same thing. In the case of our solar system there is a constant relationship between the two and this is where Newton's constant G comes from in his equation of universal gravitation. Gravity depends on aether destruction, not aether displacement. It is possible that the ratio of aether producers to aether neutral particle and aether producers changes with time. If this were to happen then the mass of the earth would stay the same but the aether destruction would decrease. If this was to happen in the earth then the mass of the earth would stay the same but the moon would move away from the earth.

Could the density of the aether around the earth ever change to such an extent that the earth moves closer to the sun? Well this is also a good question. Throughout the lifetime of the earth there have been extended periods when the earth was much hotter that it is now and times when the earth was much colder. Many theories have been put forwards for this, usually revolving about the rate that the sun is burning its fuel. There is another possible explanation. If the density of the aether is not completely uniform throughout the galaxy then the earth would move closer or further away from the sun. First realize that the aether, in relation to our galaxy, is static and that the earth and sun move through the aether at a constant speed without being affected by the aether. The sun destroys aether and as mentioned above the aether pressure gradient depends on this destruction. If the density of the aether is higher then there would be more particles in a given volume of space. Less space would be 'consumed' in order to get the required number of aether particles into the sun. Since there are more aether particles per unit volume then the speed of the aether flow into the sun would be less. The speed of the aether flowing into the sun would be lower and the earth would move further from the sun. Remember, as we demonstrated earlier, that the acceleration of the aether going past a planet toward the sun is proportional to the square root of the distance of that planet to the sun. The closer the planet is, the greater the acceleration. Conversely if we decrease the aether acceleration then the planet moves further away. Now if the density of the aether were to vary throughout our galaxy, then, as the earth moved in and out of areas of increased or decreased aether density the earth would move closer and further away from the sun. There would be no changes whatsoever to the mass of the sun or the earth and no changes to the radiation from the sun, and no changes to the aether destruction rate of the sun.

Now if there were changes to the density of the aether in some parts of the galaxy would we not detect this by its ability to bend light. Well I think this would be very hard to detect because it would be spread out over a very large area, and the aether would not be accelerating. Remember to bend light the aether has to be very significantly accelerating. We know that our galaxy is moving towards the Andromeda galaxy. If our galaxy is accelerating, which we don't really know, then there would be an aether pressure gradient across the galaxy and the aether pressure on one side of the galaxy would indeed be higher than on the other side.

Our sun is said to orbit the galaxy every 240 million years at a speed of about 220 Km/sec. If the galaxy is accelerating then the earth and all the planets would move closer and further away from the sun in a cycle with a period of 240 million years, due to the fact that the aether density would be higher on the one side of the galaxy than the other. Nir Shaviv demonstrated that the ice ages have a periodicity of 145 million years and this is also the periodicity that cosmic rays increase and decrease. There is a well defined periodicity but it is not 145 million years. Don't assume that the cosmic rays are coming from our sun. Most cosmic rays come from the center of the galaxy. This suggests something more profound. When the moon moves away for the earth, the earth moves away from the sun, and the sun moves away from the center of the galaxy. An explanation for this could be that the density of the aether in the region of space where our galaxy is fluctuates with a periodicity of 240 million years. If this theory is correct then all stars would move closer to the galactic center, all planets would move closer to their suns, and all moons would move closer to their respective planets at the same time. If the ice ages were due to solely to changes in the sun's energy output then there would be no to and fro movement of the planets and certainly not the moons of the planets, and you could not explain the closely linked fluctuation in cosmic ray radiation.

I propose that the periodicity is 145 million years because the orbit of the sun around the galaxy is 145 million years and the accepted figure of 240 million years is wrong. If the mass of the center of our galaxy is much larger than has been calculated, as my figures suggest it is, then the orbit of our sun around the galaxy would be faster. If our galaxy is accelerating, and is in a stable accelerating aether flow, then the density of the aether on the one side of the galaxy would be higher than on the other. The aether density in our region of space would then cycle with a period of 145 million years and all bodies in our solar system would be affected, in that all planets and moons would move closer and further away from each other with this periodicity. Our solar system would be affected more than other solar systems in our galaxy because we are near the edge of the galaxy. This postulate is also based on the acceleration of our galaxy being in the same plane as the plane of the galaxy. If it was at right angles then there would be no effect.

Many of the readers, if there are any left, will have their own gravity theories. If your gravity theory cannot explain the proven observation, that the moon is moving away from the earth, then you had better come up with a better theory.

Continuous creation of stars in galaxies

The next thing to think about is the continuous creation of stars in our galaxy. When stars explode heavy elements are created. I think that most people know that heavy elements are made in supernova explosions. The remains of the exploded star are spread through space. Much later these same atoms are gathered together and new stars and planets are formed out of the debris of previous stars. New star formation appears along the leading edge of the spirals in spiral galaxies from massive clouds of hydrogen gas called nebulae. I drive my car for two hundred thousand miles, and after a long and useful life, I take it to the scrap yard for recycling. It is melted down and the metal is used for making new cars. This is logical thing to do, but I also know that it does not just happen. The steel mill uses a lot of energy to recycle the steel. But, it appears that the stars in our galaxy are recycled in a similar manner but with no expenditure of energy. How is this possible? I think that the answer is that it is not possible. Energy is indeed expended to recycle the stars and that comes from aether surrounding the galaxy. Gravity pushes the hydrogen together into large clouds which eventually ignite into new stars. But there is a cost. The supply of aether is not unlimited. The intergalactic aether is gradually used up in the process, and the force of gravity is gradually weakening. This will cause the universe to expand. There appears to be basically nothing in intergalactic space except aether. If there were aether producing particles in this region then these would interfere with the transmission of light. The fact that light is transmitted with extraordinary precision over vast distances suggests that there really is nothing in intergalactic space. This is why the aether density is decreasing and the universe is expanding.

Dark matter and dark energy

Dark matter has been created to explain the rotation of the galaxies. Basically the masses of the galaxies appear to be too light to explain the observed rotations. My figures suggest that the calculated mass of the black hole at the center of our galaxy is too small by a factor of nearly 500. If you're basic gravity equation is wrong, or being applied incorrectly, then you need to add a 'fudge factor' to get the equation to work. It is not politically correct to call it a 'fudge factor' so you have to call it something with a more intelligent sounding name like 'dark matter'. Why do you need dark matter? Because the current theory of gravity does not work properly and, more importantly, you have to have something with a catchy name that will look good on new research grant applications. It is tough to make a living as an astronomer as you are always basically a full time professional beggar. To make matters worse you are dependent for your livelihood on people who haven't a clue what you are talking about. Dark matter certainly has a catchy name if nothing else.

Dark energy is another concept that has been created to fix a problem with the current gravity theory. As mentioned earlier the universe seems to be expanding. The problem is not just that it is expanding, because this could be easily explained by the big bang theory. The universe would be expanding because of the initial explosion. No, the problem is more profound. It is that the rate of expansion is increasing. This cannot be explained by the big bang. Einstein used a 'fudge factor' he call the cosmological constant to try and solve this problem. Dark energy is basically a resurrection of this idea under a different name. The idea is that, if the expansion is accelerating, then there must be source of energy for this. This is called dark energy. My theory offers another explanation. If the aether is not being continuously created in intergalactic space, then as the universe ages it will gradually be used up, and the density of the aether would gradually decrease. The force of gravity between the galaxies is due to accelerating aether flows into the galaxies. As the density of the aether decrease then the force of gravity would get weaker and the galaxies would move further apart. Not only would this happen but the rate of expansion would be accelerating, as has been observed. There would be another consequent of the decreasing aether density. The speed of light would gradually decrease. It is also possible that the frequency of light would decrease as the aether density decreases. There would be many consequences of a falling aether pressure besides the effect on gravity. All the wave like properties of light would change, including colour, diffraction, and refraction. Other forces which, as mentioned earlier, I suspect are also related to accelerating aether flow like magnetism and the strong nuclear forces would also be affected. Be that as it may, if the big bang theory is correct then the expansion of the universe would suggest that aether is not being continuously made in intergalactic space and the amount of aether we have if a finite amount and decreasing.

Hubble's law

This law states that the more distant the galaxy is the faster it is receding with a speed proportional to its distance. This value is given today as 75 km/sec per mega parsec. (per 3.262 million light years). To put this into perspective the distance to the nearest galaxy, which is the Andromeda galaxy, is about 2 million light years. This means that Andromeda should be moving away from us at the speed of about 150 Km/sec. Unfortunately this does not hold for galaxies close to ours. Galaxies exist in clusters and these interact with each other. The closest galaxies are actually moving toward us, but these are exceptions. For the majority of galaxies this equation holds. The Hubble constant has been calculated from observations of the cosmic microwave background and observations with the Hubble space telescope. This was initially calculated by Hubble using Cepheid variables which are, very importantly, independent of the red shift. Today the red shift is used to get an idea of the distance and speed of distant galaxies where Cepheid variables cannot be seen. It is nevertheless important to remember that the red shift is really the only way of calculating the speed of very distant galaxies and if there is something wrong with the red shift theory then everything else will be wrong. Even more interesting, the speed of the distant galaxies seems to be increasing. So why is the universe expanding? If gravity is simply an attraction then it should be getting smaller, or at least the expansion should be slowing down, but this does not seem to be the case. If you know the Hubble constant you can work backwards and find out when the universe began. This has been done and gives the figure of about 14 billion years ago. Any theory of gravity worth its salt must explain, or at least attempt to explain, the expansion of the universe. Einstein introduced his cosmological constant to do this. Today the concept of dark energy has been created to do this. Virtually nothing is known about this. It has been called the 5th force. The ideal universe would be neither expanding nor contracting. To get to this ideal state dark energy would have to make up 73% of the mass of the universe. How this could actually cause the expansion is rather vague.

Could there be something wrong with the red shift theory? Well yes there certainly could be. Over vast distances the properties of light might change. The speed might change or the frequency might change. According to this theory of gravity that I am proposing it is very likely that in the distant past the density of aether was different to what it is today, and this would affect the speed of light. We would never know if there was such a change because we are observing from only one point. It is indeed unfortunate that we have to but all our eggs in one red shift basket. Do I think that the red shift results are wrong? No not really, they seem to be pretty solid and an expanding universe fits my theory very well.

Olbers's paradox and the Hubble deep field photograph.

In 1823 the astronomer Heinrich Olbers pointed out, as others before and after him, had also done, that the background of the sky is black because the universe is not infinite. If the universe was infinite then there would always be some galaxy or star in our line of sight, wherever we looked, and the background would be very bright, not black. People have argued that the universe is infinite, but that the light from very distant galaxies has somehow got lost, or degenerated, on its way to the earth. If you have not really studied the Hubble deep field photograph taken by the Hubble telescope, which is reproduced in virtually every modern astronomy book, then you really should, or else you will miss out on what is probably the most significant photograph ever taken. You can see for yourself that Hoyle could well have been correct. An apparently empty region of space in Ursa Major was chosen where earth bound telescopes could not see any objects. This area was incredibly small, representing an area of space about the diameter of a tennis ball at a distance of 100 meters. In January 1996 the Hubble telescope took many exposures of the area for a period of 10 days. The composite photograph revealed at least 1500 galaxies. This picture was really a shock to the scientific community in general. It showed that the universe is vastly bigger than anyone had thought possible. I have looked at this with a magnifying glass and even the tiniest speck appears to be a galaxy. Not only are there immense numbers of galaxies, but all types of galaxies appear to be represented. There have been other deep field pictures taken since the first one. On March 9th 2004 the Hubble ultra deep field picture was shown. This was a million second exposure. It shows roughly 10,000 galaxies. It is basically the same as the first picture. What it does show very clearly, is that in the regions where there are no galaxies, the background is definitely black.

This raises another issue and that is the age of the universe. In the Hubble ultra deep field image we are looking back in time about 12 billion years. Now in this picture, it is quite clear to me, that even the tiniest galaxies have well formed spiral structures. If the big bang theory is correct then we are to assume that these galaxies are in the region of only 2-3 billion years old, which is less than the age of our own sun. Stars are continually created and destroyed in galaxies. Our own sun has been formed from the debris of previous stars and there and many generations of star ancestors making up our sun. I think it is very unlikely that the tiny perfectly formed spiral galaxies that I, and others can clearly see in the deep file pictures, are only half the age of our own sun. This just does not make sense to me. This suggests to me at least, that the universe is much older than 14.5 billion years. So what's going on? Why is the background black? If Hoyle was right, and the universe is infinite then there would be no regions of pure black because the light would have had plenty of time to reach us. The conclusion that the background is black because the universe is not infinite is not necessarily correct, and the fundamental reason lies in the reducing density of the aether.

To me the ultra deep field picture suggests that the big bang theory is wrong. I feel sure that when even more powerful telescopes are built and an 'empty' region in the Hubble deep field picture is examined in a similar manner, they will show a similar picture, but with one very significant difference. As the magnification increases the ratio of objects to the black background will decrease until eventually you will see pure black with no objects at all. This will not be due to the fact that you are looking beyond the big bang where there is nothing, but due to the recession of the galaxies, which is caused by a universal drop in aether density. I propose that our visible universe is just one small part of a much larger universe, and that our visible universe is in orbit around a gigantic central region. Let me explain this theory in some detail because it explains the increasing rate of the expansion of the universe, and solves Olbers's paradox.

The Hubble constant is 75 Km /sec per 3.262 million light years. What does this mean and what are the implications. At a distance of 7.524 million light years the speed will be 150 Km/sec and so on. At what distance will the galaxies be receding from us at a speed of light, namely 300,000 Km/sec?

This is pretty simple to work out. Make D the distance.

The equation is $D/(3.262 \times 10^{6)} = (3 \times 10^{5)} / 75$

$D = (3 \times 10^5 / 75) \times 3.262 \times 10^6$

$D = 9.786 \times 10^{11} / 75$

$D = 13.05 \times 10^9$ light years

Thus we can see that if we apply Hubble's constant, at a distance of 13 billion light years or so the galaxies will be receding at the speed of light. As you can see this is very close to the estimated 'age' of the universe. This is also the distance seen in the Hubble deep field picture.

Now why are the galaxies all moving away from us, and why is the acceleration increasing. If the big bang theory is correct then the galaxies would all be moving away from each other. But this cannot explain why the expansion is not slowing down. Indeed it appears to be accelerating, rather than slowing down. For the expansion to be accelerating then there would have to be an additional source of energy. Indeed even for the expansion to be remaining a constant there would also have to be an additional source of energy. The problem with this is that, according to the big bang theory, there cannot be any additional sources of energy, or anything else for that matter, as everything was created at one point in time.

Let me explain what I think is going on. Beyond a distance of 13 billion years the galaxies are receding from us at a speed which is faster than light. As the speed of the receding galaxies increase towards the speed of light the light is red shifted out of the visible spectrum into the infra red, and then shifted even beyond that. Beyond and adjacent to the infrared region, we get the microwave region of the spectrum. I postulate that the cosmic microwave background radiation, supposed to be the residue from the big bang, is just ordinary light that has been red shifted into the microwave region of the spectrum, just before it disappears from view forever. Do I hear howls of protest that galaxies cannot travel faster than light? Now remember what I said earlier. The speed of light is a constant because the density of the aether is a constant. The aether can flow faster than light, as it does as it enters a black hole, but the speed of light within the aether flow will still be a constant. Because the galaxies beyond 13 billion light years are travelling away from us faster than light we can never see them. This is the explanation of Olbers paradox. The sky is black despite the fact that the universe can be infinite in size.

I propose that the universe is vastly bigger than we think. All the galaxies that are within 13 billion light years of us, and that can be seen, are only a tiny fraction of a much larger universe. The universe has a dense central region and all galaxies, including our own, are in orbit around the dense central region. We can never see the center of the universe as we are receding from it faster than the speed of light.

One possible explanation for the galaxies to be moving apart is if the aether density of the universe is falling. There is a gradual fall in aether pressure, as aether is being continually destroyed within matter, but is not being replaced. The galaxies are in orbit around the center of the universe as it is destroying aether, and gravity operates here exactly as it does in our visible universe. Aether is not being created in intergalactic space, so there is a gradual fall in aether density. This causes the force of gravity to weaken and all the galaxies are slowly moving away from the center of the universe. Because they are all moving away from the center, what we see is that all galaxies are moving away from us. This explanation does not need the creation of extra energy, and explains what we see in the Hubble Deep field pictures, namely that the universe appears uniform even to the very edge. I see the universe as being similar to a galaxy, but it is a galaxy made of galaxies. The universe is vastly larger and older than we think. We will forever only be able to see a tiny fraction of it.

The other possibility is similar to the above, expect that the galaxies are flying apart because the center of the universe is destroying less aether with time, and this is weakening the aether flow towards the center, and therefore the gravity. One possible way we could tell the difference between the two possibilities is to consider the recession of the moon. If the first possibility is correct then Hubble's law will apply everywhere, because the density of aether is the same everywhere, and Hubble's constant is measuring the fall in aether density. If the expansion is due to decreasing aether destruction in the center of the universe then the intergalactic aether density would not be falling.

Hubble's constant and the recession of the moon

The first postulate I proposed is that our galaxy is moving away from the center of the universe, and all other galaxies, because of a slow drop in the aether density, and the Hubble constant is a measure of this drop in density. Since the aether density is the same everywhere, because the aether equilibrates, it is logical to propose that the speed of the moon receding from the earth is due to the same phenomenon, and this speed could be determined using the Hubble constant. This might sound farfetched, but let's try it anyway, and do the mathematics. The proposition is that if Hubble's constant applies to the moon then there is a universal fall in aether density causing the expansion of the universe. Let's do the math and see.

Don't mind the pun but let's bring Hubble's law down to earth. The constant is 75 Km/sec per 3.262 million light years. A light year is 9.46×10^{12} Km.

In order to compare the values what we need to do is bring the distance to Km and the speed to cm/year.

3.262 million light years = $3.262 \times 10^6 \times 9.46 \times 10^{12}$ Km.
This works out to 30.86×10^{18} Km

Hubble's constant is therefore $75 / 30.86 \times 10^{18} = 2.43 \times 10^{-18}$ Km/sec per Km

The distance to the moon is 407,000 Km.
The speed of regression would then be 407,000 x 2.43 x 10^{-18} Km/sec = 989,000 x 10^{-18}
= 9.89 x 10^{-13} Km/sec

Now we convert seconds to years

= 9.89 x 10^{-13} x 60 x 60 x 24 x 365.25
= 3.15 x 10^{-5} Km/year

Now we convert kilometers to centimeter. For those primitive Americans who are not too familiar with metric units there are 1000 meters in a Kilometer and 100 centimeters in a meter.

= 3.15 x 10^{-5} x 1000 x 100
= 3.15 cm per year

Now this really is amazing. From the assumption that Hubble's law applies throughout the universe, including the space between the earth and the moon, we get a calculated lunar regression rate of 3.15 cm per year. The actual measured rate is 3.9 cm per year. The closeness of these figures is absolutely astounding when you consider what we were starting from, namely figures of 10^{18} Km. The small difference could be due to a combination of inaccurate terrestrial measurement of the regression speed, and changes in the Hubble constant over time.
When we measure the recession of the moon we are using current speeds. When we look at distant galaxies we are using the Hubble constant as it was millions and even billions of years ago. Unfortunately we cannot get a good recent figure of the Hubble constant because the closest galaxies are actually moving towards us, and not receding. This is because galaxies exist in clusters and we are thought to be at the center of a galactic cluster. If gravity were to weaken over time the expansion would accelerate. We are using a Hubble constant that is out of date. However I feel that the difference is so small that this cannot be a coincidence, and this gives me a lot of confidence that the universe is actually expanding because there is a universal fall in the aether pressure.

This observation now enables us to calculate with reasonable certainty the orbit of the earth millions of years ago, as well as calculating the rate of fall in the aether density.

The regression of the earth

We all know that the earth was a lot hotter when the dinosaurs were living. What we want to know is, what was the orbit of the earth 100 million years ago, when the dinosaurs roamed the earth. How will we calculate this? The first postulate to make is that the speed of regression for the earth away from the sun will also follow Hubble's law.

The distance from the sun to the earth is 150 x 10^6 Km.
We will use the Hubble constant in the form we derived above namely 2.43 x 10^{-18} Km/sec per Km.
The distance it regresses in one second is therefore 2.43 x 10^{-18} x 150 x 10^6 Km

This works out to 364.5 x 10^{-12} = 3.645 x 10^{-10} Km in one second.

Converting this to years we get $3.645 \times 10^{-10} \times 60 \times 60 \times 24 \times 364.25$

$= 114,712,524 \times 10^{-10} = 1.14 \times 10^{-2}$ Km $= 1.14 \times 10^{1}$ m $= 11.4$ m per year.

One hundred million years ago the earth was 11.4×10^{8} m closer to the sun.

This is 1.14×10^{9} m $= 1.14$ million Km. The sun was therefore orbiting at a distance of 148.86 million Km instead of what it is today namely 150 million Km. The difference is very small. This would have undoubtedly heated up the earth but probably not to the extent that it was in the time of the dinosaurs. Four billion years ago or so, when the earth was formed the difference was much bigger. Four billion years ago it would have been $40 \times 1.14 = 45.6$ million Km closer to the sun. It would have been orbiting the sun at a distance of about 104.4 million Km.

We will be forever ignorant

My theory of the structure of the universe would indicate that the big bang theory is wrong. Since the aether pressure is falling this suggests that the aether was probably created along with all the matter that makes up the universe. But we will never know what is really going on, as we will never be able to see the entire universe. Since we are receding from the center of the universe faster than light we will never be able to see the center of the universe, where many of the solutions to the greatest mysteries of all might be found. It does not matter how gigantic telescopes become in the future, we will always remain in ignorance. This is really very depressing when you think about it. There may even be more than one universe, but that will also forever remain a speculation. When the astronauts left their reflectors on the moon I wonder if they could ever have thought what conclusions would be drawn one day from experiments performed using these items.

How to prove this aethereal theory of gravity

I am very lucky to live in a world where I don't need any money or equipment to produce a gravity theory that is at least plausible. All I need is a working brain, the ability to read, and the internet. However to prove this theory will require rather a bit more than this.

As I see it here would have to be two separate proofs of this theory. The first would be to prove that the aether exists and the second would be to prove that all objects in the universe have an excess of aether destruction over aether production. Showing that the aether accelerates into the sun or the earth would be good enough. Both of these proofs will be a tall order.

I criticized Einstein's theory of gravity, and also showed that the fundamental mistake he made when he canned the aether was not realizing the importance of acceleration. Maybe it was because he was German, and he may not have been exposed so much to Newton's law that force = mass x acceleration, as is the case in English speaking countries. If the experiment of Michelson and Morley, or a similar experiment, were reproduced on board an accelerating spacecraft then a difference in the speed of light could be demonstrated.

Lets for argument, build a spacecraft with a boom 150 meter from the craft, with a laser reflector on the end of the boom. Let's place the craft in orbit around the earth. If a laser beam is reflected from the ship to the reflector and back to the ship it would take one millionth of a second for light to travel the 300m distance to the reflector and back. Light travels 300,000,000 m/sec. Now let's accelerated the craft to 5g which is about the maximum the crew could withstand while doing this experiment. This is an acceleration of 50 m/sec^2. If a light beam was exposed to this acceleration, then in one millionth of a second it would move by 50/1,000,000 meters. This works out to 0.05mm. This is a rough and ready calculation but you can see that the reflected beam could be focused onto a photographic plate through a microscope and this difference would be enough to measure. The light would not come back to where it started from as the transmitting source is accelerating and when the light returns the source has moved. You would not have to resort to using interference patterns to demonstrate the difference, as the pioneers did, as we now have lasers and far superior optics. When the acceleration stops the beam would return to the starting point even though the craft is moving much faster. This would prove that light is bent by accelerating aether and that aether flowing at a constant speed has no effect. As the breaks are applied the beam would bend in the opposite direction. The experiment would have to be done parallel to the surface of the sun to eliminate gravity and far enough away from the earth so that earth's gravity would have no effect. The engineering problems of ensuring that the boom does not bend during the acceleration, or taking this inevitable bending into account would be massive. I think that if NASA has a few billion dollars to spare then this would be a good project. After all it is not every day that an American could upstage both Einstein and Newton. A theory remains only a theory until it is proved.

The second part of the proof would be to prove that there is an accelerating aether flow into an object. The experiment with heavy water that I discussed previously could be offered as a proof. If I am correct and the regular water hits the ground before the normal water then it would be very difficult to explain this difference by any other mechanism except the different resistance offered to the aether because of the relative differences in surface areas. Certainly according to Newton's theory they should hit the surface at exactly the same time because the surface areas of the subatomic particles play no role whatsoever in gravity.

Other gravity theories

There are many theories of gravity out there. You can look them up under crackpot.com if you want. No I am not joking; crackpot.com actually exists and is quite a good site. Some are better than others. I think mine is the best, and the only one that is actually correct, but I might be wrong. I have been wrong once or twice before. Unfortunately almost all the gravity theories are killed stone dead by mathematics. Mathematicians are pleasant enough fellows but if you allow them to operate without supervision, or if you withhold their medication, then they will come up with insanities like warped space-time, string theory and gravity waves.

In my opinion for a gravity theory to be a serious contender it has to, as a minimum, fulfill these criteria.
1. It must be able to explain why gravity is related to the square of the distance and not the cube.
2. It must be able to explain the recession of the moon.
3. It must be able to explain how orbiting planets and moons are able to break Newton's first law, and follow curved paths and not straight lines.
4. It must be able to explain all the components of the equation of Newton's law of universal gravitation.
5. It must be able to explain in detail, Galileo's observations that all objects fall at the same speed.
6. It must offer an explanation for the accelerating expansion of the universe.

7. It must propose an experiment that can actually prove the theory to be correct, even if the author cannot do the experiment himself.

I am starting the section on magnetism. The basic laws of aether flow form the basis of this force. The fundamental difference between gravity and magnetism is that gravity is due to accelerating aether putting objects into motion, whereas magnetism is due to accelerating electrically charged particles putting the aether into motion.

References

I am not going to insult your intelligence by giving you a long, boring list of selected articles, which I think will help my case, while deliberately leaving our source that might debunk, or cast doubt on my theory. Almost everything in this article is common knowledge. Where you need to look up specific facts I have given the scientists names. Don't be lazy, if you have a query just use Google and look it up.

Acknowledgement

I would like to thank my brother Ian Atkinson who first gave me the idea, 40 years ago now, that gravity could not be an attraction. He has written his own book on the cause of gravity. Being an engineer he went over the top with mathematics, and while being very interesting to read, in my opinion it is not correct. In particular his theory fails the receding moon test, as well as being unable to properly explain the increasing rate of expansion of the universe. He also offers no method of proving his theory. My work could not have been done at all without the power of the internet and Google and Wikipedia in particular. I would also like to thank my family Marilyn, Joy and Lisa for their support. I would like to thank my friend Prof. Barry Mendelow, a man with a truly huge intellect, for his positive encouragement early on when I was formulating the ideas for this theory.

www.ingramcontent.com/pod-product-compliance
Lightning Source LLC
Chambersburg PA
CBHW081221170526
45165CB00009B/2896